Analog Circuits and Signal Processing

Series Editors

Mohammed Ismail
Mohamad Sawan

For further volumes:
http://www.springer.com/series/7381

Michael Figueiredo · João Goes
Guiomar Evans

Reference-Free CMOS Pipeline Analog-to-Digital Converters

Michael Figueiredo
Centre of Technology and Systems
Universidade Nova de Lisboa
Quinta da Torre
Monte da Caparica
2829-516
Caparica
Portugal

Guiomar Evans
Departamento de Física
Faculdade de Ciências
Universidade de Lisboa
Edifício C8
1749-016
Lisboa
Portugal

João Goes
Department of Electrical Engineering
Universidade Nova de Lisboa
Quinta da Torre
Monte da Caparica
2829-516
Caparica
Portugal

ISBN 978-1-4899-8555-2 ISBN 978-1-4614-3467-2 (eBook)
DOI 10.1007/978-1-4614-3467-2
Springer New York Heidelberg Dordrecht London

Softcover reprint of the hardcover 1st edition 2013

Printed on acid-free paper

Springer is part of Springer Science+Business Media (www.springer.com)

To our families

Preface

More and more signal processing is being transferred to the digital domain to profit from the technological enhancement of digital circuits. Where technology scaling enhances the capabilities of digital circuits, it degrades the performance of analog circuits. However, it is important to note that the impact that technology scaling has on digital circuits is becoming smaller and smaller, which means that, in nanotechnologies, to enhance energy and area efficiency, we cannot simply depend on the benefits of this scaling. Although a share of the efficiency can be obtained from the technology, new circuit architectures and techniques have to be developed to really push the limits of efficiency.

In data converters, more specifically analog-to-digital converters (ADCs), a decision can be made: research energy and area efficient analog circuit techniques and architectures that cope with technological scaling issues, or design algorithms that use digital circuitry to assist the poor analog technological performance. The former option is the premise for the work developed in this book.

The work reported in this book explores various design techniques with the purpose of enhancing the power and area efficiency of building blocks mainly to be used in multiplying digital-to-analog converter-based ADCs. Therefore, novel analog techniques are developed for the three main blocks of an MDAC-based stage, namely, the flash quantizer, the amplifier, and the switched capacitor network of the MDAC. These techniques include self-biasing and inverter-based design for the flash quantizer and amplifier. Regarding the MDAC, it combines three techniques: unity feedback factor, insensitivity to capacitor mismatch, and current-mode reference shifting.

In the second part of this work, the designed amplifier is implemented and experimentally characterized demonstrating its practical feasibility and performance.

The final part of this work explores the design and implementation of a medium-low resolution high speed pipeline ADC incorporating all the developed circuits. Experimental results validate the feasibility of the techniques and demonstrate its attractiveness in terms of power dissipation and reduced area.

Acknowledgments

We would like to thank various people, who, through their support and work, improved the design of some of the circuits presented in this book and enhanced the quality of the text of some chapters.

These people, in no particular order, are: Prof. Rui Tavares, Prof. Nuno Paulino, Prof. Luis Oliveira, Prof. A. Steiger-Garção, Prof. M. Medeiros Silva, Prof. F. Barúqui, and Prof. A. Petraglia.

We would like to thank all our colleagues and friends at the research lab, especially Edinei Santin and João Ferreira.

Mr. Faustino for all the wirebonding work of all the ADC and amplifier chips.

We would also like to thank the Department of Electrical Engineering of the Faculdade de Ciências e Tecnologia of the Universidade Nova de Lisboa, CTS-UNINOVA, and the Department of Physics/CFMC of the Faculdade de Ciências of the Universidade de Lisboa, which, through projects IMPACT (PTDC/EEAELC/101421/2008), OBiS FRET (PTDC/CTM/099511/2008), and FCT/CAPES (227/09), financed the trips to all conferences, workshops, and cooperations, where some of the work developed in this book was presented.

Contents

Acronyms

A/D	Analog-to-Digital
AC	Alternating Current
ADC	Analog-to-Digital Converter
ATG	Asymmetrical Transmission Gate
BPPC	Bottom-Plate Parasitic Capacitance
CM	Common-Mode
CMFB	Common-Mode FeedBack
CMR	Common-Mode Range
CMRS	Current-Mode Reference Shifting
CMOS	Complementary Metal Oxide Semiconductor
DAC	Digital-to-Analog Converter
DC	Direct Current
DCL	Digital Correction Logic
DM	Differential-Mode
DNL	Differential NonLinearity
ENOB	Effective Number of Bits
ESD	ElectroStatic Discharge
ESL	Equivalent Series Inductance
ESR	Equivalent Series Resistance
FFT	Fast Fourier Transform
FM	Frequency Modulation
FoM	Figure-of-Merit
GA	Genetic Algorithm
GBW	Gain-BandWidth
GE	Gain Error
HS	High-Speed
IC	Integrated Circuit
INL	Integral NonLinearity
LSB	Least Significant Bit
MBTA	Multiply-By-Two Amplifier
MC	Monte Carlo

MDAC	Multiplying Digital-to-Analog Converter
MIM	Metal-Insulator-Metal
MOM	Metal-Oxide-Metal
MOS	Metal Oxide Semiconductor
MS	Mean-Square
MSB	Most Significant Bit
NMOS	N-channel Metal Oxide Semiconductor
OPAMP	Operational Amplifier
OS	Output Swing
OTA	Operational Transconductance Amplifier
PCB	Printed Circuit Board
PFBL	Positive FeedBack Latch
PM	Phase Margin
PMOS	P-channel Metal Oxide Semiconductor
PSD	Power Spectral Density
PVT	Process, Voltage, and Temperature
PVTM	Process, Voltage, Temperature, and Mismatch
RAM	Random Access Memory
RE	Reference Error
RF	Radio-Frequency
RS	Reference Shifting
S/H	Sample-and-Hold
SPRS	Sampling Phase Reference Shifting
SC	Switched Capacitor
SFDR	Spurious-Free Dynamic Range
SHA	Sample-and-Hold Amplifier
SNDR	Signal-to-Noise-and-Distortion Ratio
SNR	Signal-to-Noise Ratio
SR	Slew Rate
SRAM	Static Random Access Memory
TF	Transfer Function
TI	Time-Interleaved
TPPC	Top-Plate Parasitic Capacitance
THD	Total Harmonic Distortion

Chapter 1
Introduction

1.1 Motivation

The world and the signals that exist in it are inherently analog. Processing these signals in the analog domain is too complex, therefore, it is necessary to transform or convert them into a digital form. Unlike analog circuits, digital circuits are more robust to variations, are more configurable and programmable, and are easier to design. Consequently, analog-to-digital converters (ADCs) are indispensable components in systems where an interface between analog signals and their digital representation is necessary.

There is an extensive list of devices and applications where ADCs carry out key functions. These applications range from medical imaging systems to biomedical devices, from communications to instrumentation and from sensors to modern technological gadgetry. In the case of high-speed medium-low resolution (6–8 bits) ADCs, like the one presented in this book, these are widely used in wireline and wireless communications [6, 85, 89, 101], in the read channel of optical and magnetic storage devices (e.g., DVD and Blue-ray systems) [100, 120, 178], and in low cost test instrumentation (oscilloscopes) [3, 110]. The steady increase of technological consumerism and interest in novel technological conceptions such as cloud computing, cognitive and software defined radios, video-on-demand, etc., has demanded that gadgets perform better, last longer, and be smaller, i.e., the consumer has driven the design of integrated circuits (ICs), where ADCs are included, to be energy and area efficient.

As already mentioned, more and more signal processing is being pushed to the digital domain to profit from the technological enhancement of digital circuits. This is a natural outcóme due to the fact that technology scaling is largely driven by, and therefore, benefits digital circuits. Where technology scaling enhances the capabilities of digital circuits, it degrades the performance of analog circuits. Note however that, the positive impact that technology scaling has on digital circuits is becoming smaller and smaller [17]. Voltage supply and parasitic capacitances are not scaling the way they used to. This means that, in nanotechnologies, to enhance energy and

M. Figueiredo et al., *Reference-Free CMOS Pipeline Analog-to-Digital Converters*,
Analog Circuits and Signal Processing, DOI: 10.1007/978-1-4614-3467-2_1,

area efficiency, we can not simply depend on the benefits of technology scaling and digital circuits. Although, a share of the efficiency can be obtained from the technology (this part comes almost for free), the major efficiency gain factor will come from the development of new architectures and algorithms, and computer-aided design tools.

What concerns the implementation and development of ADCs, a decision can be made: search and research energy and area efficient analog circuit techniques and architectures that cope with technological scaling issues, or design algorithms that use digital circuitry to assist the poor analog technological performance. The former option is the premise for the work developed in this book.

Therefore, the research goals and objectives of this work focus on the development of analog design techniques and architectures that enhance the energy and area efficiency of MDAC-based ADC topologies. In order to achieve this, these techniques should:

- Enhance the performance of the ADC's building blocks, namely, the comparator, the amplifier, the multiplying-DAC, and the reference voltage circuitry.
- Cope with modern technological issues such as process, supply voltage, and temperature (PVT) variability.
- The circuits should be designed in a standard digital CMOS technologies without using analog options that increase the cost of the design.

All circuits, if possible, should be demonstrated in integrated silicon prototypes to assess their functionality and performance.

1.2 Original Contributions

The main contributions of the work described in this book, based on work proposed in previous publications, extend from the development of a flash quantizer to an operational transconductance amplifier (OTA) and from techniques for multiplying-DACs (MDACs) to a complete ADC. These contributions have lead to the production of various papers in conferences and journals of the area. The contributions of this work can be described as (in chronological order):

- Design of a 1.5-bit flash quantizer with built-in switching threshold levels using two techniques [48]. The first based on analog inverters (similar to threshold inverter quantization) and the second technique based on self-biasing. The latter improves robustness against PVT variations. Analog inverters combine the pre-amplifier and latch of the conventional flash quantizer, and permit building-in the switching threshold levels. Employing these two techniques allowed the design of a highly efficient inverter-based self-biased 1.5-bit flash quantizer with built-in switching thresholds.
- Design of a highly efficient two-stage OTA based on two techniques: self-biasing and inverters [47, 50]. Again self-biasing is employed to improve robustness

against PVT variations and to eliminate the biasing circuitry overhead. This technique allowed achieving an approximately constant open-loop DC gain. The second technique based on inverters allowed increasing the amplifier's efficiency in terms of power-to-speed ratio. By using inverters the total transconductance per unit current is effectively doubled. Employing these two techniques allowed the design of a two-stage inverter-based self-biased CMOS OTA with improved efficiency. An IC prototype of the OTA was developed to validate its performance.
- Design of a unity feedback factor multiply-by-two amplifier (MBTA) [49] which ultimately lead to the design of two unity feedback factor 1.5-bit MDACs. One of the MDACs performs reference shifting during the sampling phase to maintain a unity feedback factor during the amplification phase, while the other performs reference shifting in current-mode during the amplification phase. The technique employed to obtain the gain of two in the MBTA and MDAC circuits is sensitive to parasitic capacitors, which is minimized with a simple analog compensation technique. The developed unity feedback factor configuration allows effectively doubling the energy efficiency of MDACs.
- The main contribution of this work is the design of a 7-bit 1 GS/s two-way interleaved pipeline CMOS ADC [51]. The designed ADC combines all previously mentioned blocks: the inverter-based self-biased 1.5-bit flash quantizer, the two-stage inverter-based self-biased OTA, and the unity feedback factor 1.5-bit MDAC with current-mode reference shifting. Given that the flash quantizer has built-in thresholds and the MDAC performs reference shifting in current-mode, the developed ADC does not need any reference voltages and, thus, precludes reference voltage circuitry. An IC prototype was implemented to validate the performance and functionality of the techniques employed to design the ADC.

1.3 Book Organization

The book is organized into 7 chapters. The present chapter gives an introduction and defines the motivation behind the work presented in this book.

Chapter 2 provides a general background for all the building blocks described throughout the book. MDAC-based ADC architectures are briefly described, which includes their advantages and limitations. Special focus will be given to the building blocks of pipeline ADCs, where besides detailing the function and importance of each block, related errors and performance limiting aspects will also be given. Various static and dynamic performance parameters, and metrics that characterise ADCs are described. Finally, the chapter ends with a state-of-the-art of medium-low resolution high-speed MDAC-based ADCs, as well as, surveys of two-stage amplifiers and reference voltage circuits (in the context of A/D conversion).

Chapter 3 describes the two proposed capacitor-mismatch insensitive MDAC architectures with unity feedback factor. Besides describing the basic concept, various analyses are carried out, namely, gain error, reference shifting error, feedback factor, and noise. These detailed analyses show the benefits and limitations of each

MDAC circuit. To conclude the section, a comparison is carried out with other mismatch-insensitive MDACs and MBTAs.

Chapter 4 is divided into two major sections. In the first section, the proposed 1.5-bit flash quantizer is described. Various analyses are carried out, namely, kickback noise, regeneration time, metastability, offset, sensitivity to common-mode variations, and finally, the results from a working proof of a pipeline ADC that employs the proposed circuit in all stages is shown. The section concludes with a design procedure and a comparison with other comparator circuits is carried out. The second section presents the proposed two-stage OTA. The analyses carried out include differential-mode and common-mode feedback, noise, offset, slew rate, input-output ranges, and some considerations are given what concerns the amplifier's class. Finally, guidelines are given for a successful design and a genetic algorithm optimization procedure is briefly shown.

In Chap. 5 the design of a 7-bit 1 GS/s time-interleaved pipeline ADC is discussed. All the building blocks used in the design of the ADC are thoroughly described.

Chapter 6 discusses the implementation of two IC prototypes with the objective of evaluating their performance and functionality. The first circuit concerns the proposed two-stage OTA and the second prototype concerns the 7-bit 1 GS/s time-interleaved pipeline ADC. For each prototype, a floorplan, a layout, design considerations, a printed circuit board (PCB), and a test setup will be given. Each section will end with the demonstration of the experimental results achieved and a comparison will be carried out with the state-of-the-art presented in Chap. 2.

Chapter 7 draws the main conclusions of the work carried out in this book, summarizing all the employed techniques and developed building blocks. A section dedicated to suggestions for future work is also given.

Finally, the book concludes with an appendix that discusses a possible solution for on-chip integration of the reference shifting currents.

Chapter 2
General Overview of Pipeline Analog-to-Digital Converters

Abstract This chapter provides a general background for the work carried out in this book. Therefore, its purpose is to cover all aspects of the developed work. First, some A/D converter (ADC) architectures will be briefly described. The common element of these architectures is the use of the multiplying-DAC (MDAC) circuit as their principal block. Advantages and limitations of the architectures will also be given. The MDAC circuit is one of the key elements of this book. Given that this work presents a prototype of a pipeline ADC, it is important to describe each of its building blocks. Besides detailing the function and importance of each block, related errors and performance limiting aspects will also be given. After the description of the pipeline converter sub-blocks, various static and dynamic performance parameters, and metrics that characterise ADCs are given. It will be the objective here to explain the parameters that fundamentally dictate the performance of ADCs. Finally, the chapter is completed with a state-of-the-art of medium-low resolution high-speed pipeline ADCs. Besides this overview, surveys of two key building blocks, namely, two-stage amplifiers and reference voltage circuits (in the context of A/D conversion), which deserved special attention in this work, are also presented.

2.1 MDAC-Based Analog-to-Digital Converter Architectures

There are many architectures of A/D converters, each with their own set of characteristics and capabilities to be used in different applications. Well known architectures are Full-Flash (or Parallel), Two-Step, Sub-Ranging, Folding, Integrating, Successive Approximation (SA), Algorithmic, Pipeline, Sigma-Delta modulators, and Time-to-Digital. It is also possible to find numerous combinations of the various existing topologies, such as: time-interleaving can be used as a means of increasing the sampling frequency (conversion rate) by arranging various converters of the same type in parallel; it is very frequent to find interpolation associated with flash and folding converters; the two-step topology can employ either flash or SA architecture in each

M. Figueiredo et al., *Reference-Free CMOS Pipeline Analog-to-Digital Converters*, Analog Circuits and Signal Processing, DOI: 10.1007/978-1-4614-3467-2_2,

step; the pipeline and algorithmic topologies usually employ flash converters in each step, etc.

Part of the work carried out in this book implements an ADC topology that employs MDAC circuits, consequently, the only converter topologies of interest, of the ones mentioned above, are those that use MDAC circuits as a means of obtaining a residue with amplification (i.e., simultaneous DAC, subtraction, and residue amplification functions). With this in mind the only architectures that will be discussed are the Two-Step (flash), the multi-step Algorithmic, and the Pipeline, targeting higher conversion rates. The time-interleaving technique will also be briefly described.

2.1.1 Two-Step Flash ADC

The Two-Step Flash architecture evolved from the Full-Flash converter. One of the main drawbacks of the latter is the number of necessary comparators, given by $N_{comp.} = 2^N - 1$, which scales exponentially with the resolution of the converter (N), making it, in some cases, impractical to implement due to the necessary die area. The Two-Step Flash topology alleviates the number of necessary comparators by quantizing the input in two steps, hence its name, as shown in Fig. 2.1a. The effective reduction factor in the number of comparators when compared to the Full-Flash ADC, is exponentially proportional to the converter's resolution, and is approximately given by[1] $2^{\frac{N}{2}-1}$. In other words, the higher the resolution the more area efficient it becomes to use a Two-Step topology.

As shown in Fig. 2.1a, each step (or stage) is composed of a quantizer, with a resolution (N_1 and N_2) inferior to the resolution (N) of the entire converter, thus requiring less reference voltages and comparators, and consequently, occupying less die area. Between the two steps an amplified residue voltage needs to be generated, which is achieved with a DAC, a subtraction operation block and a gain block. These three blocks constitute an MDAC circuit. The principle of operation is as follows: the input is sampled by the first quantizer during the sampling phase. During the residue amplification phase, the first quantizer decides the most significant bits (MSBs), which are then used to reconstruct a voltage (using the DAC), that is subtracted from the original sampled input and then amplified (by 2^{N_1}) to create an amplified residue voltage. Still during this phase, the second quantizer samples this residue. The objective of the amplification is to restore the residue to the full voltage range of the converter, thus facilitating the implementation of the second quantizer (Fig. 2.1b), or eventually, reusing the same quantizer in a cyclic way. During the final phase, the second quantizer (of resolution N_2) quantizes the residue to obtain the least significant bits (LSBs). The final digital output is assembled using digital logic, by adding the MSBs together with the LSBs.

Basically, this topology simplifies the quantization, by trading comparators with time. The Full-Flash converter achieves a quantization in two clock phases (one clock cycle), while the Two-Step needs at least three phases, i.e., one and a half clock

[1] Assuming that the same number of bits is extracted in both quantization steps, i.e. $N_1 = N_2 = N/2$.

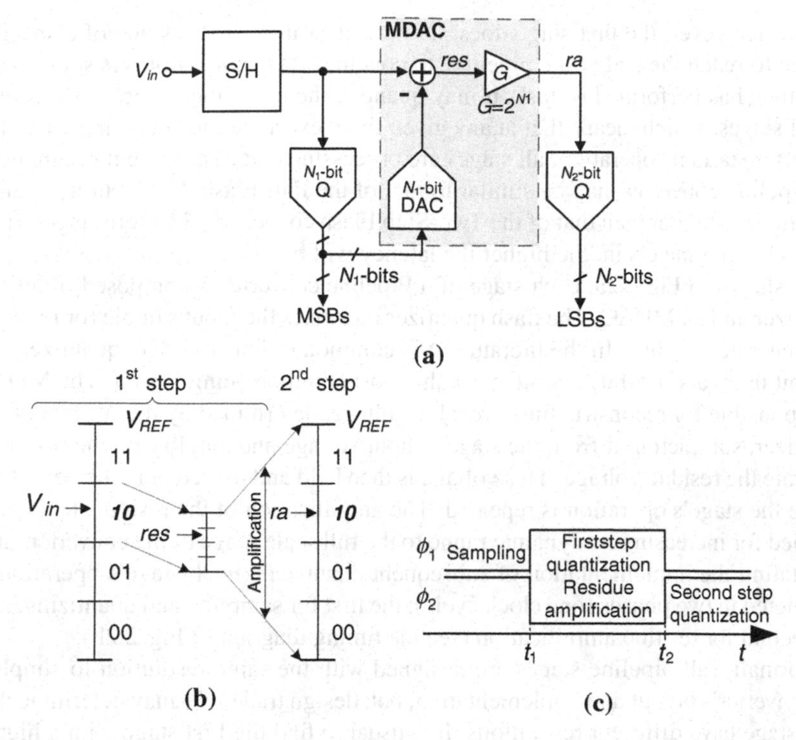

Fig. 2.1 The Two-Step A/D converter: **a** Block diagram. **b** Example of residue amplification. **c** Timing diagram

cycles (Fig. 2.1c). Although the throughput may be similar to that of the Full-Flash converter (one digital output per clock cycle), the Two-Step has higher latency.[2] If N_1 is made equal to N_2, then only one quantizer needs to be designed and, therefore, both quantizers may use the same reference voltages. The latter is also made possible by using a residue amplification gain of 2^{N_1}. The above assumptions consider that no digital redundancy is used.

If more steps (or stages) are added to the converter to simplify its implementation and relax the requirements of each step, which would eventually lead to a minimum resolution per step ($N = 1$), the resulting A/D architecture would be the Pipeline.

2.1.2 Pipeline ADC

The Pipeline converter's operation is basically the same as that of the Two-Step Flash. Each stage is responsible for quantizing N_j-bits, $j = 1, 2, \ldots, K$ ($N_j < N$) and generating an amplified residue for further quantization (performed by subsequent

[2] Latency is the number of initial clock cycles to produce the first digital output.

stages). However, the first stage does not have to wait for the residue of a specific sample to reach the end of the pipeline, to conclude its quantization. As soon as the first stage has performed its task, it may quantize the next input sample. This holds for all stages, which means that at any given time, except at the beginning when the converter starts its operation, all stages are processing data. Thus, the throughput of the Pipeline converter may be similar to that of the Full-Flash ADC, but its latency is high, even higher than that of the Two-Step Flash converter. The more stages there are in the pipeline chain, the higher the latency will be.

As shown in Fig. 2.2a, each stage of a Pipeline converter is composed of a flash quantizer and an MDAC. The flash quantizer quantizes the input sample (or residue) and generates N_j bits. In the literature it is common to find 1 to 4 bit quantizers (in half-bit intervals, 1.5-bit, 2.5-bit, etc.), the most common being 1.5-bit. The MDAC is responsible for reconstructing a residue voltage, determined by the N_j bits of the quantizer, subtracting it from the stage's input voltage and amplifying the result, to generate the residue voltage. This voltage is then held and passed onto the next stage where the stage's operation is repeated. The amplification of the residue, by 2^{N_j}, is justified for increasing its dynamic range to the full scale range of the converter, thus facilitating the implementation of subsequent quantizers. Each stage's operation is completed in two phases (one clock cycle): the first for sampling and quantizing, and the second for residue amplification (see the timing diagram of Fig. 2.2b).

Normally, all pipeline stages are designed with the same resolution to simplify the converter's layout and implementation, but, design trade-offs may determine that each stage have different resolutions. It is usual to find the first stage with a higher resolution. Another important reason for all stages to be equal (in resolution) is that the reference voltages are the same for all quantizers and DAC functions.

What concerns digital logic, the Pipeline converter employs digital synchronization logic to align (over time) all bits before producing the final digital output word. Aside from synchronizing, these converters usually employ digital correction, which corrects for nonidealities in the flash quantizers [97]. The Two-Step Flash converter, described above, may also employ digital correction logic. These digital blocks are detailed further on in this chapter.

2.1.3 Multi-Step Algorithmic ADC

The Algorithmic (or Cyclic) converter, as the name indicates, quantizes the input sample in an algorithmic or repetitive manner. Its principle of operation is basically the same as that of the Pipeline converter, in that an input is sampled, quantized, and a residue is amplified, and then the quantization and residue amplification process is repeated by subsequent stages. The main difference with the Algorithmic converter is that the conversion algorithm (sampling, quantization, and residue amplification) is repeated in the same physical space (reutilizing the same circuits) or area, while the Pipeline ADC repeats its operation over more area. In other words, the Algorithmic converter trades space for time, which means it has a longer conversion cycle.

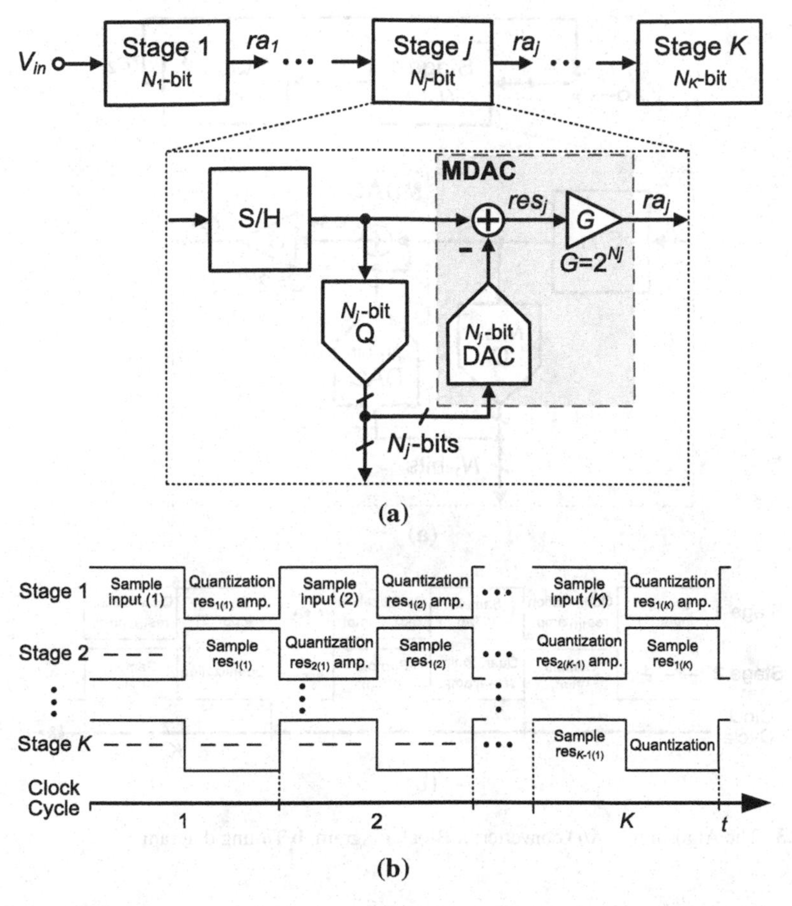

Fig. 2.2 The Pipeline A/D converter: **a** Block diagram of a stage, exemplifying its building blocks. **b** Timing diagram

The converter has a minimum number of stages (usually two), where the residue repeatedly passes through each stage, successively generating output bits, from the MSB to the LSB, as shown in Fig. 2.3a. After the LSB is generated, the process starts over again with a new sample.

The principle of operation is as follows: the input voltage is sampled by the first stage. It is then quantized to generate the MSBs. These bits are used to reconstruct a voltage (using a DAC and reference voltages) that is subtracted from the input voltage generating the residue voltage. This residue is then amplified (and held) to the full scale range of the converter and sampled by the second stage. This process repeats itself between the first and second stages until the LSB is generated. At this point a full digital output is ready, while the converter is sampling the next input sample (Fig. 2.3b).

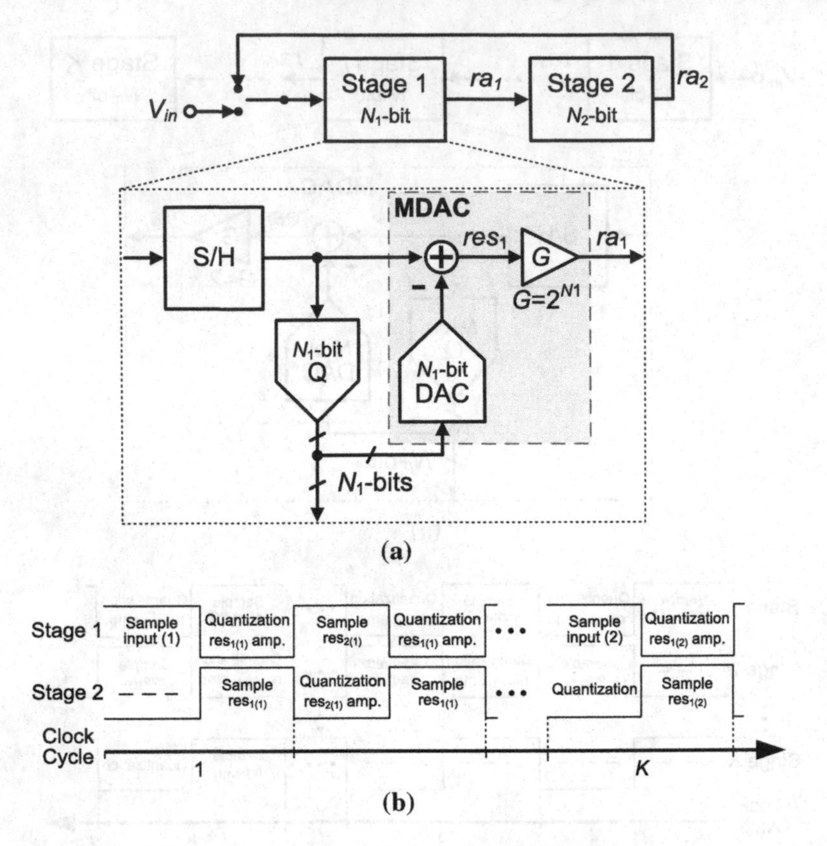

Fig. 2.3 The Algorithmic A/D converter: **a** Block diagram. **b** Timing diagram

Unlike the Pipeline, this converter is unable to process more than one sample at any given time, thus a converter of resolution N needs $N+1$ clock cycles to quantize a single input sample. Compared to the Pipeline ADC, this converter's throughput is much lower, but has the same latency. It trades throughput with die area and facilitates layout and implementation due to the number of necessary blocks, which are much less than that used in a Pipeline converter.

The Algorithmic architecture must employ digital circuitry to hold and align the digital outputs of each stage before producing the final output word, and it usually employs digital correction logic.

In order to increase the throughput, the time-interleaved technique may be employed.

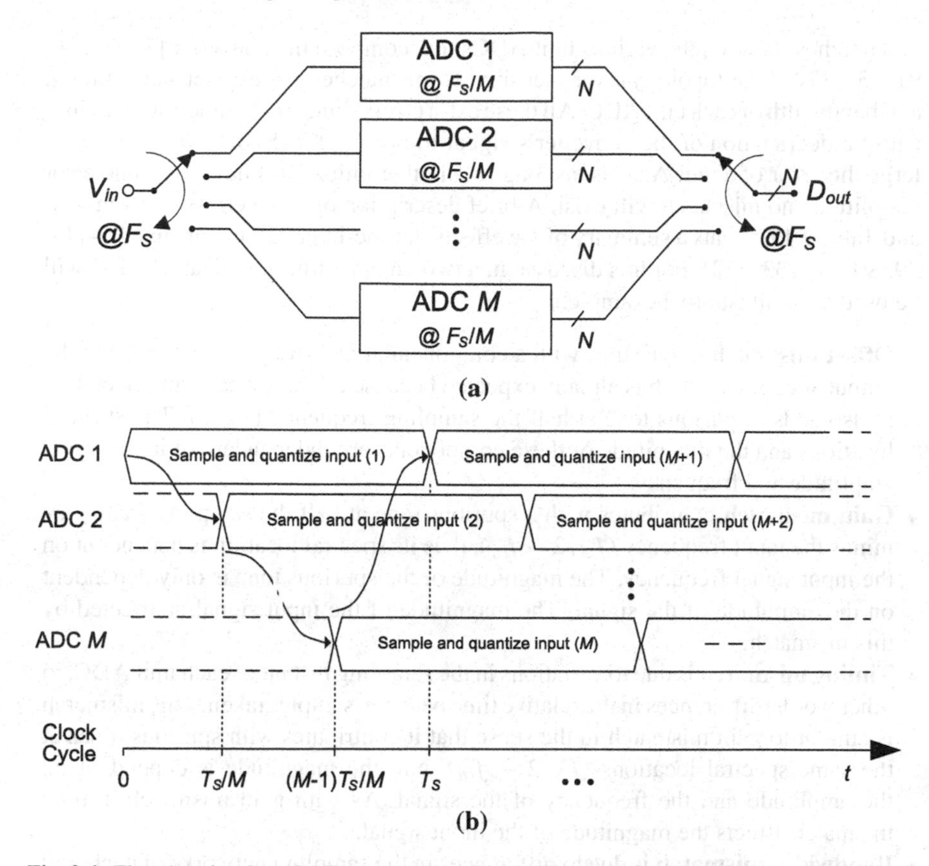

Fig. 2.4 Time-interleaving of A/D converters: **a** Block diagram. **b** Timing diagram

2.1.4 Time-Interleaving ADCs

Time-Interleaving [11] is a technique used to increase the throughput or conversion rate of a converter. This technique may be applied to all A/D converter topologies. It consists of using an array of M parallel converters multiplexed at the input and at the output as shown in Fig. 2.4a. Each converter operates at a conversion rate $F_{S/M}$ (where F_S is total conversion rate), making it easier to implement. The analog input multiplexer, adequately timed, is responsible for attributing an input sample to each of the converters over time, when it reaches the last converter in the array it starts over again. The digital output multiplexer guarantees that the timing sequence of digital outputs are in accordance with the sampled inputs. Each converter added to the array inevitably increases the die area and the power consumption of the overall A/D conversion system. The timing diagram of operation is shown in Fig. 2.4b.

Besides the limitations produced by each unit (multiplexed) ADC, the time-interleaved technique introduces its own limitations. These have mainly to do with mismatches between the various unit ADCs that compose the converter [11, 39, 80, 91, 133, 172]. This topology is very sensitive to mismatches in the offset, gain, timing, and bandwidth of each unit ADC. All these extra errors (inherent to time-interleaving) cause a degradation of the converter's signal-to-noise ratio (SNR). No matter how large the error of a unit ADC is, as long as all other unit ADCs have the same error magnitude, no mismatch will exist. A brief description of these errors is given next and Table 2.1 presents a summary of the effects of time-interleaving mismatches [11, 39, 80, 91, 133, 172]. For this discussion, a two-channel time-interleaved ADC will be used as an illustrative example.

- **Offset mismatch** contributes with a component at DC (frequency, $f = 0$) in the output spectrum, which is already expected because all ADCs have an offset, but, it also adds a spurious tone at half the sampling frequency ($F_S/2$). The spectral locations and the magnitude of these components are independent of input signal amplitude and frequency.
- **Gain mismatch** contributes with a spurious tone at half the sampling frequency minus the input frequency ($F_S/2 - f_{in}$), thus its spectral location is dependent on the input signal frequency. The magnitude of the spurious tone is only dependent on the amplitude of the signal. The magnitude of the input signal is affected by this mismatch.
- **Timing mismatch** is due to variations in the sampling instant of each unit ADC, in other words, differences in the relative time between samples taken. This mismatch is similar to gain mismatch in the sense that it contributes with spurious tones at the same spectral locations ($F_S/2 - f_{in}$), but, the magnitude is dependent on the amplitude and the frequency of the signal. As with gain mismatch, timing mismatch affects the magnitude of the input signal.
- **Bandwidth mismatch** is due to differences in the sampling networks of each unit ADC. If each unit ADC has a dedicated sample-and-hold (S/H), and if each S/H has a different bandwidth, then bandwidth mismatch will affect the performance of the overall time-interleaved ADC [39, 91]. This mismatch adds similar contributions to that of gain and timing mismatches. The main differences are that the gain part of the bandwidth mismatch is now dependent on the input signal frequency and the timing part has a nonlinear dependency with the input signal frequency.

2.2 Building Blocks of Pipeline Analog-to-Digital Converters

The objective of this section is to briefly overview each of the constituent blocks of a Pipeline A/D converter. Besides an overview, errors related to each block will also be given. Various references are given throughout the section for a more detailed coverage and further reading. Note that, not all the blocks described below are necessary to build a Pipeline ADC. For example, the sample-and-hold (S/H) and decimation blocks are not strictly necessary. It should be equally noted that the blocks described

Table 2.1 Magnitude and spectral location of spurious tones due to mismatches and their effect on the signal ($A_{in}\sin(2\pi f_{in})$) component for a two-channel time-interleaved ADC. The offset and gain mismatches are given by o_i and g_i respectively ($i = 1, 2$). The relative timing mismatch is given by $r_i = \Delta t_i / T_S$, where Δt_i is the absolute timing mismatch and $T_S (= 1/F_S)$ is the sampling period

Mismatch type	Signal component (Magnitude)	Spurious component (Spectral location) (Magnitude)	
Offset	—	DC $\frac{o_1+o_2}{2}$	$F_S/2$ $\frac{o_1-o_2}{2}$
Gain	$A_{in}\frac{g_1+g_2}{4}$	$F_S/2 - f_{in}$ $A_{in}\frac{g_1-g_2}{4}$	
Timing	$A_{in}\cos(2\pi f_{in}\frac{r_1-r_2}{2})$	$F_S/2 - f_{in}$ $A_{in}\sin(2\pi f_{in}\frac{r_2-r_1}{2})$	
Bandwidth	see [91]	$F_S/2 - f_{in}$ see [91]	

below are generic, in the sense that they are found in most Pipeline ADCs. In recent developments some of these blocks have been substituted for more efficient ones and some have been eliminated (mostly for power and/or area savings).

2.2.1 *Sample-and-Hold*

The sample-and-hold (S/H) block is found at the very beginning of the converter. Its objective is to discretise, in time, i.e., to sample the input and hold the sampled input for the subsequent block to process it. The S/H converts a continuous-time signal into a discrete-time signal (the signal is still continuous in amplitude). Another circuit with similar functions is a track-and-hold (T/H), where the main difference to a S/H is that the output of the T/H tracks (follows) the input, then samples, and finally holds the sample. A simple version of a S/H and a T/H are shown in Fig. 2.5 with their respective timing diagram.

S/H circuits operate in two phases, the sampling and the holding phase, as shown in Fig. 2.5a. During the sampling phase, the switch (ϕ_S) is closed and the capacitor is charged to the input voltage. When the switch is opened, the input is sampled, and because the charge on the capacitor can not be destroyed, the sampled voltage is held on the capacitor. At this moment, the held voltage can only be sensed by a high input impedance block such as an amplifier (unity-buffer in this case).

There are numerous errors associated to sampling and holding an input signal [9, 84, 135]. Some are mentioned below:

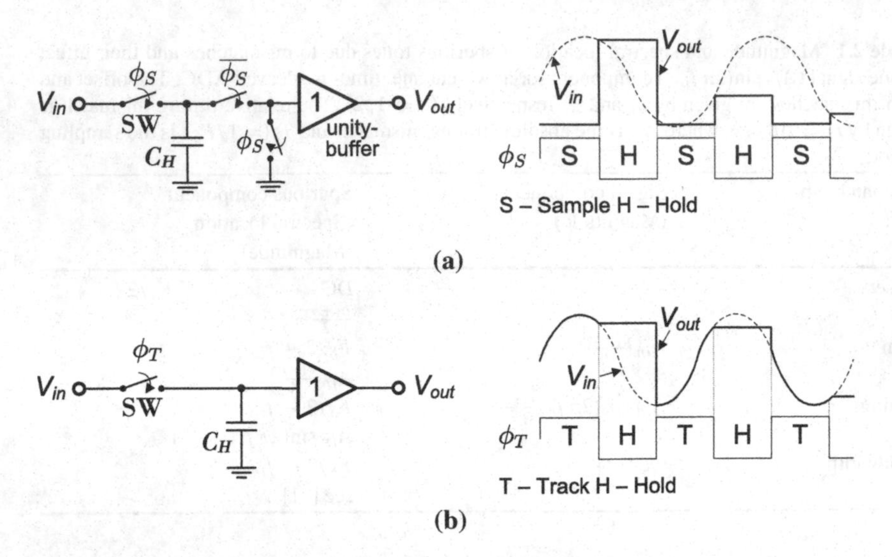

Fig. 2.5 Simple versions of (**a**) S/H and (**b**) T/H, with output buffer and output waveform

- **Finite sampling bandwidth** if the input signal's frequency is higher than the sampling bandwidth ($f_{-3\,\text{dB}} = 1/(2\pi R_{SW} C_H)$), an output signal voltage with a phase difference is sampled.
- **Acquisition time** is the time it takes the amplifier (buffer) to settle. This error is associated with amplifiers which will be explained further on.
- **Sampling uncertainty (aperture error)** is the time uncertainty at the moment of sampling. This error has two possible origins: a long rise or fall time, or a sampling instant that changes from period to period. This error is particularly problematic in the presence of high frequency signals.
- **Sampling pedestal** is the error voltage added to the sampled voltage caused by the switch while it is turning off. This extra voltage is due to channel charge injection and clock feed-through. This error is particularly problematic when the error voltage is signal dependent, which adds distortion.

2.2.2 Multiplying-DAC

As seen in the previous section, the MDAC is a circuit which performs numerous functions. These functions include sampling the input signal (or residue voltage from a previous stage), reconstructing a voltage using a DAC, obtaining a residue (subtraction of the reconstructed voltage from the stage's sampled voltage), performing a gain to amplify the residue, and finally holding the amplified residue for the next stage. The block diagram of a generic MDAC is shown Fig. 2.6. Normally, a switched-capacitor (SC) network is employed to accomplish all these functions. Sampling is

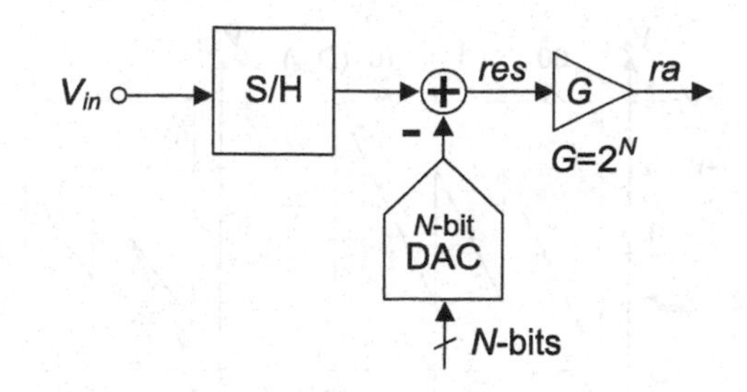

Fig. 2.6 Block diagram of a generic N-bit MDAC

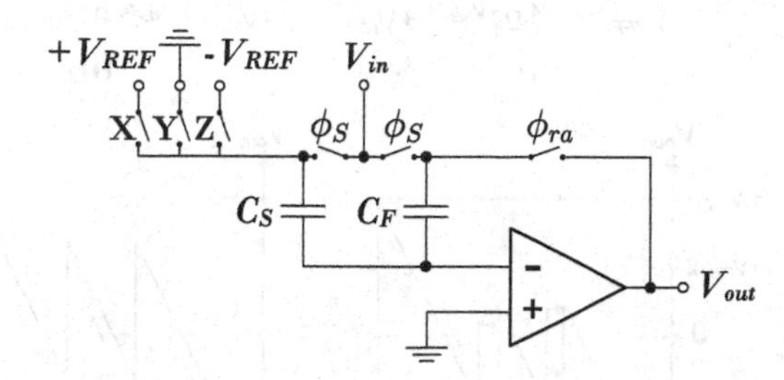

Fig. 2.7 Closed-loop switched-capacitor opamp-based 1.5-bit MDAC

achieved by means of switches and capacitors, similar to that shown in the previous subsection. The MDAC employs an operational amplifier (opamp) with a capacitive feedback network (closed-loop) to provide the DAC, subtraction, and amplification functions.

In the literature it is possible to find various techniques of implementing the MDAC function, either in open or closed-loop, such as, closed-loop switched-capacitor techniques, switched-current techniques [103], open-loop amplification [114], dynamic source follower amplification [76], MOS parametric amplification [123], and the substitution of the opamp for a comparator based circuit [15], among others. Each technique has its own advantages and limitations, not discussed here.

To understand the principle of operation of an MDAC circuit, a simple closed-loop switched-capacitor opamp-based 1.5-bit MDAC is used, as shown Fig. 2.7 (single-ended shown for simplicity) [98]. This resolution MDAC (1.5-bit) is chosen because it is one of the most widely used of all implemented stage resolutions. The input-output transfer characteristic is shown in Fig. 2.8a (ideal case). This characteristic can be described by the following expression

$$V_{out} = 2V_{in} + B \cdot V_{REF}, \tag{2.1}$$

where B represents the bit decisions made by the local quantizer (represented in Fig. 2.7 by X, Y, or Z which represent $B = +1$, $B = 0$, and $B = -1$, respectively) and V_{REF} represents the converters reference voltage. Equation 2.1 shows that the MDAC has a gain component ($2V_{in}$) and a reference shifting component ($B \cdot V_{REF}$).

Circuit operation is as follows: during ϕ_S the input (V_{in}) is sampled onto capacitors C_S and C_F. During the residue amplification phase (ϕ_{ra}), C_F is put in the opamp's feedback loop, and, due to charge conservation, the charge on C_S is transferred to

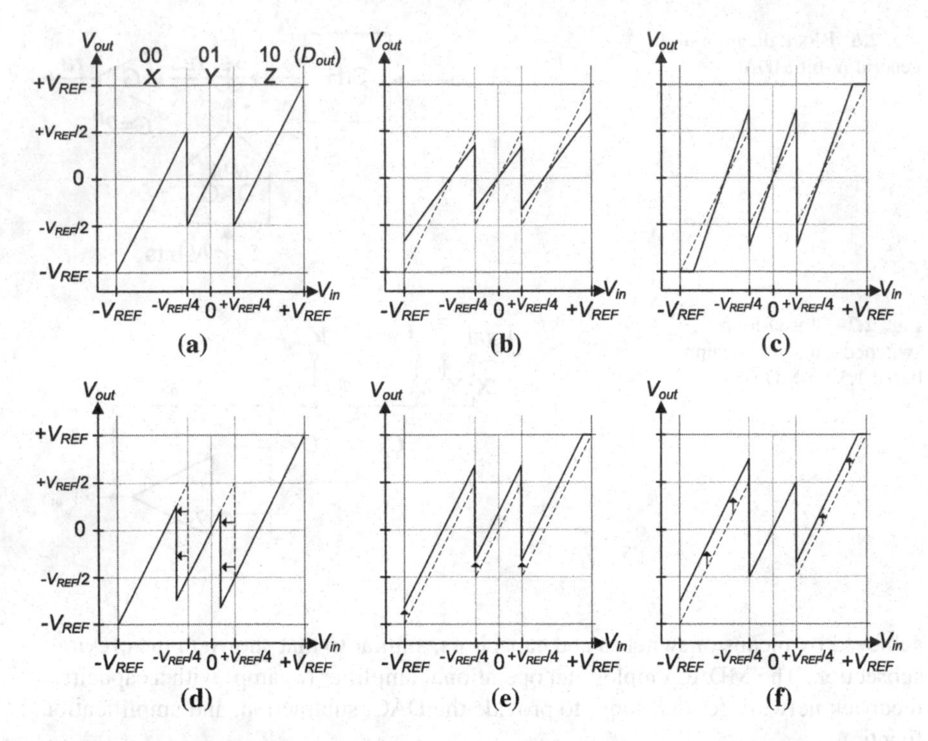

Fig. 2.8 Input–Output characteristics of the 1.5-bit MDAC and the effects of errors (*dashed line* represents the ideal situation): **a** Ideal. **b** Gain error: gain < 2. **c** Gain error: gain > 2. **d** Offset of the quantizer decision levels. **e** Offset in the MDAC due to charge injection or offset of the opamp. **f** DAC nonlinearity

C_F. The amount of transferred charge depends on the quantizer's decision (X, Y, or Z). If Y is high, i.e., if the Y switch is closed, the sampled input charge on C_S is transferred to C_F. In this case the output (V_{out}) will simply be $2V_{in}$ (parameter B of Eq. 2.1 is 0). If X or Z is enabled, then there will respectively be a charge addition or subtraction on C_S and the resultant charge transferred to C_F. In this single-ended version example, the DAC is only composed of capacitor C_S.

The limiting factors of closed-loop SC-MDAC circuits are given next. To aid the enumeration of these factors, the example of the 1.5-bit MDAC will be used as well as its transfer characteristics shown in Fig. 2.8.

- **Gain error** caused by capacitor mismatch (between C_S and C_F) and finite opamp DC gain [97, 102, 156]. Slopes of the characteristic vary from the ideal value of 2. See Fig. 2.8b, c.
- **Offset errors** can be caused by offset of the quantizer decision levels (Fig. 2.8d), by charge injection [86], or by opamp offset in the MDAC (Fig. 2.8e). Offset errors are of minor importance given that most errors can be corrected by the digital correction logic [97].

- **Nonlinearity errors** caused by DAC capacitor mismatch and nonlinearity errors present in the opamp [102] (Fig. 2.8f).
- **Thermal noise** comes from the ON-resistance of the switches which is sampled on the sampling capacitors and from the opamp [156].
- **Speed** conversion rate is limited by the opamp's closed-loop configuration, namely, the opamp's finite speed (GBW) and the closed-loop feedback factor [98] (slew rate may also be a limiting factor of the speed).

2.2.3 Local Flash Quantizer and Comparators

As the name indicates, this circuit is a quantizer based on the Full-Flash converter topology. In Fig. 2.9a, a generic N-bit flash quantizer is shown. It employs, in parallel, a number of comparators,[3] each with their own reference voltage, to be compared with the input signal. The output of each comparator is either 0 or 1, indicating that the input signal's voltage is lower or higher than the reference voltage, respectively. There are various methods of implementing the comparator: cascade of inverters (or simple gain stages), an opamp in open-loop, and the latched comparator [66, 84]. The first two circuits are designed to amplify the input signal or the difference between the input and reference signals to guarantee a logic output (0 or 1, or, in terms of voltage, the negative or positive saturation voltage, respectively). The latched comparator is composed of a pre-amplifier and a positive feedback latch. The pre-amplifier amplifies the small differential input signal and minimizes effects caused by the latch, while the latch guarantees logic levels at the comparator's output.

In Fig. 2.9b, the last block of the 1.5-bit flash quantizer is an XYZ encoder. This circuit is responsible for guaranteeing, depending on the decisions of the comparators, that either X, Y, or Z is one. As mentioned before, these signals decide the B parameter of the reference shifting of the MDAC (see Eq. 2.1). Output bits (b_i, b_j) are used for digital correction (discussed further on).

The reference voltages may be generated by a resistive or capacitive divider string, or by means of a SC network at the comparators' input. Normally for an N-bit flash quantizer, $2^N - 1$ comparators are necessary. For a half-bit quantizer (1.5-bit, 2.5-bit, etc.), only $2^N - 2$ comparators are necessary. N represents the nearest integer greater than the half-bit resolution, e.g., 1.5-bit corresponds to $N = 2$.

Factors that limit the performance of comparators and, thus, quantizers are given below:

- **Offset** the opamp and the pre-amplifier introduce an input-referred offset error which may be due to mismatches (between circuit components) or may be inherent to the comparator design [52, 84, 140].
- **Charge injection and clock feed-through** caused by channel charge and parasitics associated with the sampling switches while turning off [84, 135].

[3] The comparator is probably the most widely used component in A/D conversion, fundamental in practically all topologies.

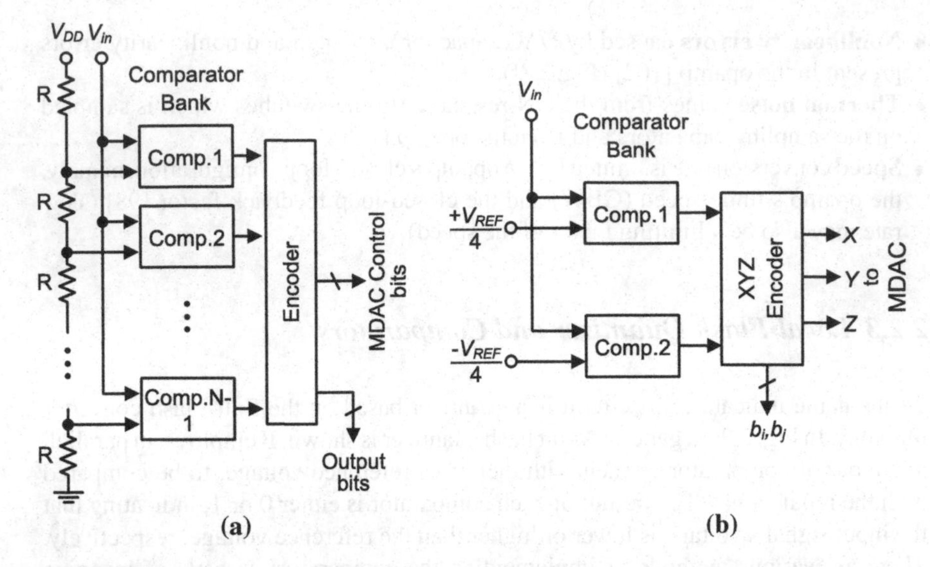

Fig. 2.9 Flash quantizer examples: **a** Generic N-bit quantizer. **b** 1.5-bit quantizer

- **Kickback noise** this error occurs during switching in latched comparators. When the comparator goes into latch mode, the high speed of the positive feedback latch causes high speed transients, which inject charge through parasitic capacitors back into the input signal, thus causing unwanted disturbances [52, 84, 140].
- **Comparison time** this is the time the comparator takes to produce a valid digital output. The worst case scenario is described by an overdrive recovery test, which defines the time the comparator takes to recover from a large input immediately followed by a very small input [104, 140].
- **Metastability** occurs when a very small input renders the comparator to be unable to produce a valid digital output. This error may be given as a probability of the occurrence of a metastable state [105, 140], as a number of metastable states per second [135], or, for the case of a flash quantizer with various comparators, as the mean time between failures (taking into account the number of comparators) [1].

2.2.4 *Operational Amplifier and Common-Mode Feedback Circuitry*

The operational amplifier, better known as opamp, is probably the most important and most used block in analog signal processing. The word operational comes from the fact that these blocks can be used to implement various functional operations. It is an active circuit which ideally has high gain, high input impedance, and low output impedance. Typically working in the voltage domain, a voltage is inputted and an

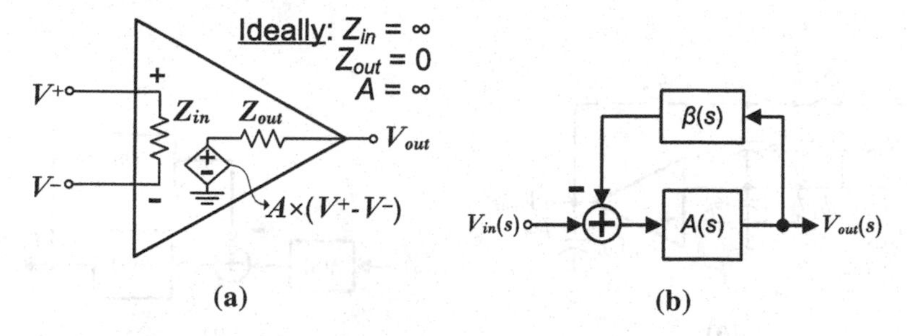

Fig. 2.10 Opamp: **a** Basic symbol. **b** Simplified block diagram of a feedback circuit

amplified voltage is provided at the output, i.e., the opamp is a voltage-controlled voltage source ($V_{out} = A \times (V^+ - V^-)$, where A is the opamp's open-loop gain), as shown in Fig. 2.10a. There are more types of amplifiers, but the one that deserves attention, given the work developed in this book, is the operational transconductance amplifier (OTA). This type of amplifier achieves a high open-loop gain (at low frequencies) at the expense of a high output impedance. They are widely used in SC circuits and do not need low output impedance because they usually only drive pure capacitive loads (and not resistive loads). Their gain and speed depend on the transconductance of specific transistors that compose the OTA.

Opamps usually operate in closed-loop form, inheriting all the associated benefits such as, less sensitivity to circuit and process, supply voltage, and temperature (PVT) variations, thus less distortion, higher input impedance, lower output impedance, and higher bandwidth. Figure 2.10b depicts a simplified block diagram of a negative closed-loop system, where $A(s)$ represents the opamp's open-loop gain (it represents Fig. 2.10a), $\beta(s)$ is the feedback network (with associated feedback factor), and s is the complex frequency.

As an example of the functionality and importance of the opamp, the SC circuit of Fig. 2.11a is used (this circuit is similar to Fig. 2.7 for $Y = 1$). For this example C_L represents a load capacitor and the objective is to double the input voltage (i.e., $V_{out} = 2\,V_{in}$). During ϕ_1, V_{in} is sampled onto C_S and C_F. During ϕ_2, the opamp (in closed-loop now) forces a virtual ground at its inverting input and, hence the voltage across C_S to zero. Due to charge conservation, the charge stored on C_S, $Q_S = C_S V_{in}$, is transferred to C_F. No charge goes to the amplifier because of its high input impedance. If C_F equals C_S then V_{out} becomes $2V_{in}$. Another way of exemplifying the importance of the opamp is: during ϕ_1, the voltages at the top and bottom plates of C_S are V_{in} and zero, respectively. Immediately at the beginning of ϕ_2, the top plate of C_S is connected to zero, which forces its bottom plate (and the opamp's inverting input) to $-V_{in}$. Simultaneously, the opamp (in closed-loop now) forces the virtual ground at its inverting input. In order to accomplish the virtual ground, the output voltage has to be increased by the same amount of voltage, i.e., V_{in}. Finally, the output will become $2V_{in}$ at the end of ϕ_2. If no opamp is used in the aforementioned explanations, neither charge conservation nor virtual ground would

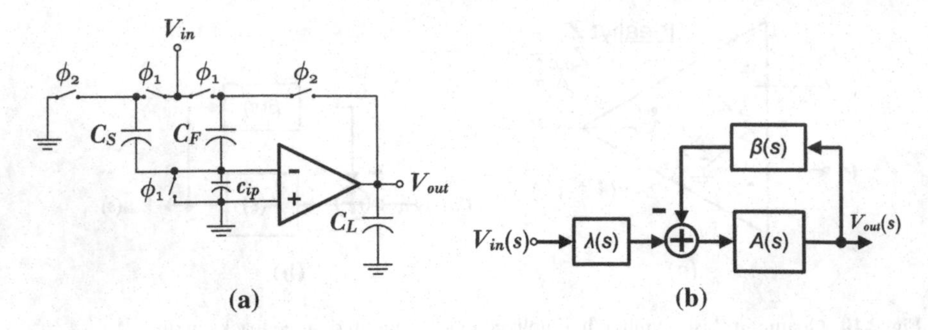

Fig. 2.11 Example using an opamp in a feedback loop: **a** SC circuit. **b** Simplified block diagram of the feedback scheme

occur. Consequently, the sampled charge would simply be distributed between C_S, C_F, and C_L, but V_{out} would not be $2V_{in}$ at the end of ϕ_2.

Figure 2.11b illustrates the equivalent block diagram of the SC circuit of Fig. 2.11a, where the feedback network is given by $\beta(s) = C_F/(C_F + C_S + c_{ip})$ and the input network is $\lambda(s) = (C_S + C_F)/(C_F + C_S + c_{ip})$ [28].

At the transistor level, many amplifier architectures exist. It will not be the objective of this work to describe any, but rather to give an idea of the common blocks used in most of them. The basic blocks are an input stage (where the input signal is connected), a gain or differential to single-ended conversion stage, and finally, an output driver stage (used to drive the load connected at the output) [66]. It is possible that, besides the input, the output stages be differential, thus inheriting the advantages of fully differential circuits, which will be given further on. When supply voltages were high (higher than 1.8 V in older CMOS technology nodes), gain could be achieved by cascoding (stacking) transistors. In this case enough gain was achieved with a single-stage amplifier, which also has the highest speed of operation. However, due to the power reduction necessity and low supply voltage (1.2 V and lower) tendency of modern nanoscale CMOS technologies (0.13 μm and beyond), it is no longer possible to cascode transistors, therefore, gain can only be achieved by multiple stages, i.e., by cascading stages. Besides gain, the output swing also decreases due to the supply voltage reduction, thus another reason to cascade stages. At this moment, we have arrived at a two-stage amplifier: the first stage mainly for gain and the second for output swing and speed with a small contribution to the overall gain. Due to compensation (to stabilize the amplifier), which is inevitable, the speed of the two-stage amplifier is reduced when compared to its single-stage counterpart. If more gain is needed, another stage can be added (three-stage amplifier), at the expense of speed and stability issues.

There are a number of parameters that characterize the performance of opamps. Most of these are limiting factors of their performance. Depending on the application the opamp is inserted in, some become more relevant than others. Here most of them will be described [65, 84, 92, 104, 141]. The parameters are:

- **Low-frequency gain**: There are two types of gains, differential-mode (DM) and common-mode (CM) gain. Usually an opamp is sized for a given DM gain while minimizing the CM gain. The former is fundamental as it determines the precision of the overall system where the opamp is used (usually in closed-loop). These gains, measured with the opamp in open-loop, are a function of frequency. At low frequencies they are called, the DC gain. An important sub-parameter of gain is its nonlinearity for different output voltages. Normally, for a small output voltage swing, the gain has its maximum value, but for larger voltage swings, the gain tends to be smaller.
- **Bandwidth**: There are two types of bandwidths, small-signal and large-signal (discussed in the following item), which characterize the high frequency performance of opamps. The former is related to the frequency dependence of the open-loop gain, which is caused by parasitic, compensation, and/or load capacitors present in the opamp. The open-loop gain versus frequency, has a constant value for low frequencies (DC gain, explained before) and then at a certain point in the frequency, the gain starts to decrease or roll-off (at a -20 dB/decade rate). This point is the opamp's bandwidth or the location of the dominant pole. The point at which the gain reaches unity is the unity gain bandwidth (f_u) or the gain-bandwidth product (GBW). Note that, GBW = f_u is only applicable either for a single pole opamp or for an adequately compensated multi-pole opamp [92, 104]. These two design parameters, f_u and GBW, determine the speed of the system the opamp is inserted in.
- **Slew rate** or large-signal bandwidth determines the rate at which the opamp can change the output voltage in the presence of large input signals. In large signal conditions, the opamp will try to provide current to charge the capacitors (compensation, load, etc.) of the system. The rate at which it does this is called the slew rate (SR). If the current is insufficient or the opamp does not have enough time (case of SC circuits) to charge the capacitors, the output will not reach the desired value and nonlinear distortion will arise.
- **Compensation and Phase Margin**: Opamps inserted in feedback loops can be potentially unstable, if not adequately compensated. A measure of this instability is called phase margin (PM). In single-stage opamps (which are less prone to instability), compensation is normally achieved (almost for free) by the load capacitor. In multi-stage opamps (unstable by nature), compensation capacitors and compensation schemes are inevitable.
- **Settling time**: This defines the time it takes the output to reach its final value within a given settling error (associated with the desired accuracy) when a step input is applied. This is probably the most important opamp parameter for SC circuits given that it is a time domain parameter, that includes the effect of gain (and its nonlinearity), small-signal bandwidth, slew rate, phase margin, and the closed-loop's feedback factor. Figure 2.12 [162] depicts an opamp's output response to an input step showing where each mentioned parameter plays its role. If any of them are not designed correctly, their effects will be pronounced in the output response and a longer settling time will probably occur.

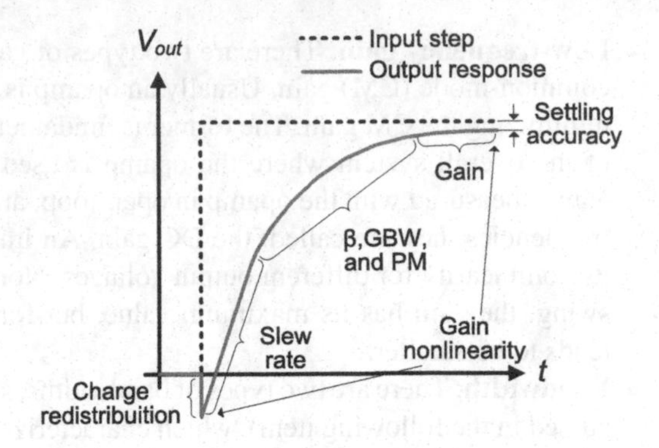

Fig. 2.12 Time-domain opamp output response (to an input step) and the role played by some opamp performance parameters

- **Output swing** is the maximum output voltage range possible that maintains the opamp functioning nominally, i.e., with the desired gain.
- **Common-mode input range** is the maximum input voltage range that guarantees negligible degradation of the opamp's performance.
- **Offset** is the output voltage when the input is zero. Ideally the output voltage should be zero, but will not be due to inherent design issues of the opamp (systematic offset) and mismatches between otherwise matched transistors (random offset). This offset voltage, if large enough, can limit the output swing.
- **Noise** is generated by devices with resistive components such as transistors (resistive channel) and, naturally, resistors. There are many types of noise, the two most commonly discussed are flicker (or $1/f$) and thermal noise. The latter is of more importance in high-speed or wideband opamps. Just as in the case of offset, the opamp may be designed for low noise but never for zero noise. Noise determines the smallest detectable input, the opamp may process.

Most parameters can be enhanced by using a fully differential structure for the opamp. In fact, opamp performance can double. This type of structure indicates that the opamp, besides having a differential input, also has a differential output. The advantages of the fully differential architecture are well known and are: larger gain, larger output swing, higher immunity to extrinsic noise, reduced distortion (suppression of even harmonics) and enhanced speed/power ratio [150]. Although the intrinsic noise level of fully differential circuits is higher, the signal-to-noise ratio (SNR) will still be higher than single-ended output circuits due to the larger output swing obtained.

An opamp designed with a fully differential structure needs to employ a special circuit to measure and control its output common-mode voltage. This circuit is known as the common-mode feedback (CMFB) circuit. Without this circuit the uncontrolled output CM voltage could drift to the supply rails. The CMFB circuit guarantees that this voltage stays approximately at midway between the supply rails. There are continuous-time and SC–CMFB circuits. Usually in SC circuits, the SC version of CMFB circuits is preferred.

2.2.5 Reference V/I and Buffering

Reference circuits are essential in analog and data converter systems. They generate reference voltages and currents that are used to bias circuits, to compare with other signals, for addition and subtraction operations, among others. In the specific case of data converters, reference circuits are determinant in defining the input and output full-scale ranges. Therefore it is necessary to guarantee a sufficient level of accuracy,[4] so that the overall performance of the data converter is not limited. To achieve this, they need to be independent of external conditions such as, process, supply voltage, temperature, and load disturbances. In the case where a reference voltage needs to drive a large capacitor or various capacitors amounting to a large capacitance (like in DAC circuits), or be used in a high-speed or high-accuracy SC circuit, an additional block needs to be added to the output of the reference circuit. Commonly known as a reference buffer, this block is used to maintain the reference voltage constant and to guarantee that it charges and discharges the capacitors it drives, in the available amount of time (particular case of SC circuits). In other words, the buffer must settle the reference voltage to within a given error, within a given time slot, which depends on the accuracy and speed of the converter.

There are many forms of generating a reference voltage (resistive or capacitive ladders, bandgap circuits, etc.), but this is not the objective of this subsection. Instead, an overview of the issues and difficulties designers have to overcome to buffer and/or stabilize a reference voltage for SC converters, will be given.

It is possible to find many forms of reference voltage schemes in the literature. The options divide into on- and off-chip buffering with (or without) the use of on- and off-chip damping resistors and decoupling capacitors. The following table describes and analyses the most used forms of reference voltage circuitry and buffering schemes.

The conclusions extracted from Table 2.2 can be summarized as follows: the reference circuitry will occupy a large area, will dissipate a large amount of power, and/or will need at least one extra pin. Most of the currently employed solutions suffer from a combination of these drawbacks. From a system-level perspective, neither system-on-chip nor system-in-package designs are benefitted from voltage-domain reference circuitry. To avoid extra costs, no extra pins can be used, therefore, on-chip high-speed buffering must be employed, a trade between cost and power is made. If extra pins are available, then off-chip decoupling may be used (with a low bandwidth buffer), but then a penalty in area is paid for the off-chip decoupling capacitors. All this to avoid on-chip decoupling because silicon area is more expensive (valuable) than discrete components.

---

[4] In [113] it is shown that the reference circuit can have 1-bit lower accuracy than the resolution of the ADC.

Table 2.2 Description and analysis of the advantages and disadvantages of different reference voltage schemes

Description	Analysis	Diagram	Performance
On-chip buffer without decoupling [12, 14, 81, 158]	• High-speed buffer with wide bandwidth. • High power consumption. • Noise performance hard to achieve (as frequency rises, buffer output impedance rises). • [81] uses a deglitch circuit to minimize glitches during switching. • Due to large current peaks, the buffer may require dedicated supply pins.	off-chip ← → on-chip Free Pad/Pin Buffer to ADC V_{REF} Ref. circuit	• Accuracy + • Power – • Speed + • On-chip area +– • Off-chip area +
On-chip buffer with internal decoupling [20, 34, 78]	• Wirebond inductance too large for off-chip decoupling due to impractical amount of ringing on reference voltage. • Large on-chip capacitors, occupying large area, are unavoidable [20]. • On-chip RC filter [34].	off-chip ← → on-chip Free Pad/Pin Buffer to ADC C_{int} Ref. circuit	• Accuracy + • Power +– • Speed +– • On-chip area – • Off-chip area +

(Continued)

Table 2.2 (Continued)

Description	Analysis	Diagram	Performance
On-chip buffer with external decoupling [22, 31, 69, 109, 155]	• Low bandwidth buffer. • Low power consumption. • Noise performance and output impedance dependent on quality (ESL and ESR) of decoupling capacitors. • Wirebond inductance causes ringing of the internally generated reference voltage. Dampen ringing with large on-chip capacitors and resistors (these occupy large area). • Reduce inductance with special packaging [69]. • Additional pins.	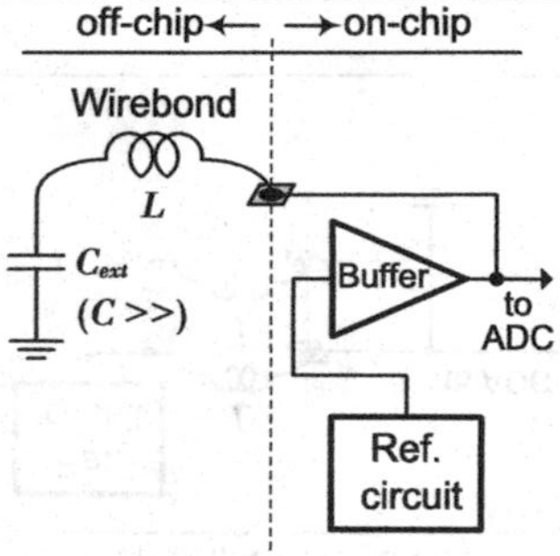 off-chip← →on-chip Wirebond L C_{ext} (C>>) Buffer to ADC Ref. circuit	• Accuracy +– • Power + • Speed – • On-chip area +– • Off-chip area –
On-chip reference without buffer with external decoupling [68]	• Reference is taken off-chip, RC filtered and dampened and brought on-chip again. • Two wirebonds in reference voltage path. • Special packaging unavoidable. • Two additional pins.	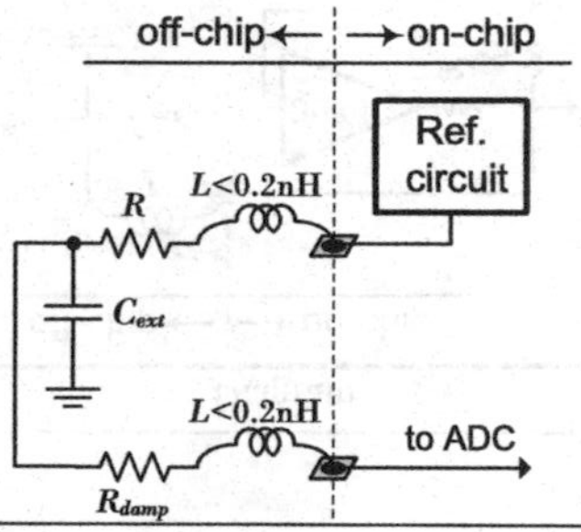 off-chip← →on-chip Ref. circuit R L<0.2nH C_{ext} L<0.2nH to ADC R_{damp}	• Accuracy +– • Power + • Speed – • On-chip area + • Off-chip area –

(Continued)

Table 2.2 (Continued)

Description	Analysis	Diagram	Performance
On-chip buffer with external and internal decoupling [117]	• Use of external and internal capacitors and damping resistors. • Special internal decoupling scheme (low-V_T decoupling capacitors) [117]. • Additional pins.	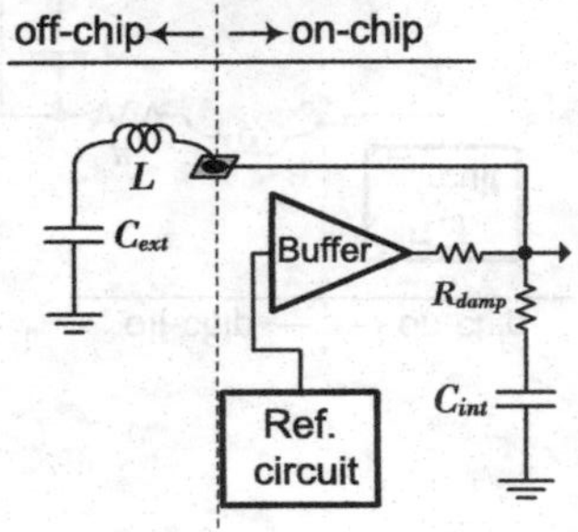 off-chip ← → on-chip L C_{ext} Buffer R_{damp} C_{int} Ref. circuit	• Accuracy +– • Power + • Speed +– • On-chip area – • Off-chip area –
Off-chip reference with internal decoupling [21, 88, 171]	• Use large on-chip capacitors to dampen the ringing caused by wirebond (occupying large area). • Internal decoupling capacitor may be larger than converter itself [171]. • For lower inductance more pins must be used [21]. • Additional pins.	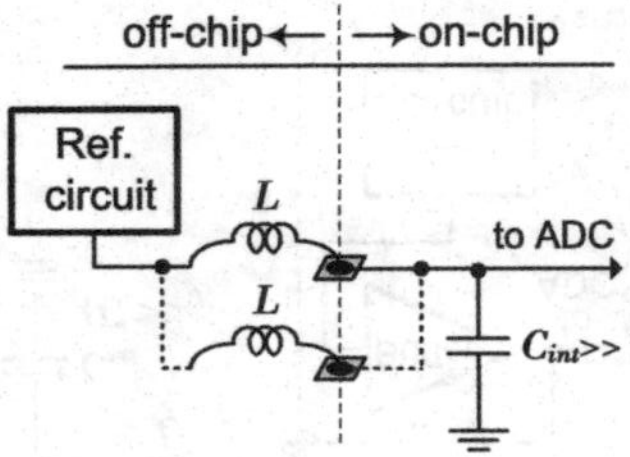 off-chip ← → on-chip Ref. circuit L L to ADC C_{int}>>	• Accuracy – • Power + • Speed – • On-chip area – • Off-chip area –

Special packaging: chip-scale flip-chip with < 0.2 nH wirebond inductances

Parameters of reference circuits and buffers that may affect the overall performance of data converters are given next. They are:

- **Ringing** is caused by the inductance (L) of the wirebond and converter's capacitance (C) during switching in SC circuits. The ringing occurs at the natural frequency of the LC circuit. To reduce and dampen the ringing, large on- and/or off-chip decoupling capacitors and damping resistors must be used [53, 88, 113, 171].
- **Speed** of a reference circuit or buffer is determinant in the presence of perturbations [141]. During clocking of the capacitors in SC circuits, the reference voltage is constantly disturbed, and has to recover before the end of the phase to avoid incomplete reference settling which causes offset errors. This error may limit the converter's conversion rate.
- **Output impedance**: It is fundamental that the output impedance of the buffer circuit be as low as possible, to adequately feed the converter with the reference voltage, avoiding large voltage drops. To achieve low output impedance at high frequencies, large capacitors need to be used, which consequently occupy a large amount of area. Nevertheless, large capacitors help reduce the noise bandwidth and suppress external disturbances, if large enough [141].
- **Noise**: Directly couples to the stage's input signal during sampling and also to the output signals during the amplification phase. Because noise limits the smallest signal that may be converted, it directly influences the full-scale range of the converter [113].
- **PSR**: Power supply rejection determines the capacity of the buffer to reject noise of the supply (V_{DD} and ground) lines from coupling to the output buffered voltage. If noise couples to the output voltage, it contributes to the total output noise of the buffer.
- **Offset** errors limit the full-scale range of the converter. Negative offsets (lower reference voltage) saturate the residue voltage, thus reducing the conversion range, which causes distortion. Positive offsets (higher reference voltage) cause residue voltages to be smaller thus degrading the signal-to-noise ratio (SNR) of the converter [113]. These offset errors cause interstage gain errors [155] and overall converter gain errors.
- **Signal dependency**: When a reference voltage and input signal are connected to the capacitor's plates, any variation in one signal causes a modulation in the other. This is called signal-dependent modulation. Therefore, if a reference voltage does not settle adequately, it induces a variation in the input signal, which is then sampled. This modulation degrades the performance of the converter [113, 155].
- **Reference distribution** is not an easy task, particularly when references must be provided to widely separated locations across the die [142], or in converters with high clock rates or high resolutions [68]. The long lines cause voltage drops which alter the reference's original value. If each pipeline stage uses a reference voltage with a different value, the performance of the converter is highly degraded.

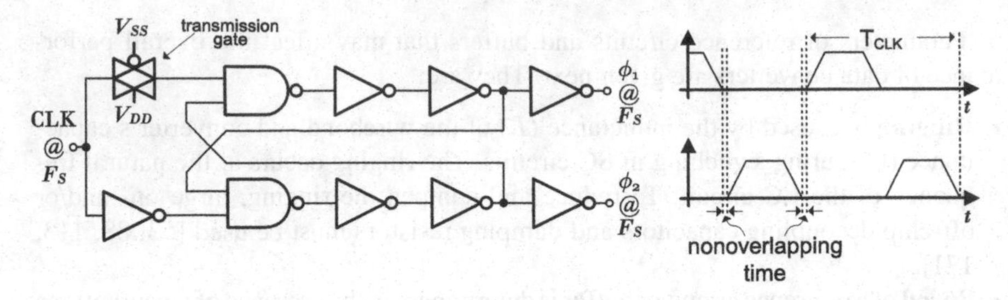

Fig. 2.13 Simple two-phase nonoverlapping clock generator and output waveforms

2.2.6 Clock Generation

Switched-capacitor circuits need clocking schemes to turn on/off its switches to achieve a specific function. Most data converters rely on a special scheme that permits the signal acquired during the sampling phase to be transferred and amplified without loss. In other words, it is imperative that no charge is destroyed or lost between the sampling and amplification phases. To achieve this, the overlapping time between phases must be zero. This is accomplished by a nonoverlapping clock generator. An example of a simple two-phase nonoverlapping clock generator, widely found and used in the literature, is shown in Fig. 2.13 [105]. This simple configuration has one clock input and two phases with 180° phase shift are provided at the output. The nonoverlapping time is controlled by the delay of the input NAND gate and the two inverters (before the feedback). To increase this time, more inverters (in an even number) can be added.

There are other schemes of generating nonoverlapped phases, namely using phase-locked loops [82] or delay-locked loops [37]. These circuits are much more complex than the simple two-phase nonoverlapping clock generator of Fig. 2.13. They are used in many applications such as in communication, data and clock recovery, FM demodulation, and high-speed or high-resolution data converters, among others. Given that they were not used in the work developed in this book, they will not be further discussed.

Clock generation or clock synthesis circuits limit the performance of data converters due to PVT variations, but mainly due to device and delay mismatches. These circuits contribute the following errors:

- **Clock jitter** is the random variation of the time the clock falls to zero (the falling edge is used here to define the sampling instant) around the ideal time. This error was briefly covered in the S/H subsection. The jittering of the timebase translates into an increase of the noise floor over all frequencies, and consequently, degrades the SNR of the converter. As mentioned before, this error is input signal dependent.
- **Clock skew** is a fundamental limitation inherent to time-interleaved converters. Each unit converter of a time-interleaved topology must sample the input at equally spaced instants in time, given by the M/F_S, where M is the number of unit converters and F_S is the total sampling frequency. Any deviation from these ideally

spaced sampling instants causes timing errors, which translate into deterministic spurious tones (visible in the converter's output spectrum), ultimately degrading the SNR of the converter. This error is also input signal dependent and was covered in Sect. 2.1.4 (timing mismatch).

2.2.7 Digital Backend and Decimation

As mentioned in Sect. 2.1 for the various topologies, a backend only composed of digital circuits is necessary. These circuits store and align the digital outputs from each stage, which then move on to the correction stage, and are finally buffered to produce the final digital output word. Basically three digital blocks are necessary: synchronization logic, digital correction logic, and output buffers. The latter block (which is basically an even number of cascaded inverters) is optional and depends on where the digital outputs need to go next and if they need to be buffered or not. To exemplify the functionality of the circuits described in this subsection, a simple 4-bit Pipeline converter with two 1.5-bit stages and a final 2-bit stage will be used.

Synchronization logic is fundamental in converters that are unable to produce a digital word in one clock cycle, i.e., are unable to digitize the analog input at once (like Full-Flash converters). This logic is mainly composed of memory and shift circuits, such as flip-flops (FF) or shift registers. Taking the example of Fig. 2.14, when the analog input signal is processed by the first stage, its comparator produces digital outputs which need to be stored because the residue voltage will only be processed by the last stage of the converter at the end of one more cycle. Therefore, during this clock cycle, the bits from the first stage need to be stored. The same analysis can be made for the outputs of the second stage, which need to be stored for half a clock cycle. As can be seen in Fig. 2.14a, FF_a and FF_b synchronize the digital outputs, while FF_{c-e} align the outputs of a specific input sample for later processing. Figure 2.14b shows the timing diagram of operations and time-alignment, i.e., where the digital outputs of each of the sampled inputs are aligned at a specific instance in time. After all digital outputs are time-aligned they are ready for digital correction.

Digital correction logic is used to correct nonidealities and indecisions in the comparators used in the quantizers [97, 98]. Therefore, the quantizer's offsets can be as large as $\pm 1/2 V_{LSB} = V_{REF}/4$ for a 1.5-bit stage, which relaxes the specifications of the quantizer. This correction logic works on the basis that each stage has redundancy which is used for correction and eliminated by the correction logic itself. For example, a 1.5-bit stage has a true resolution of 1-bit and 0.5-bit redundancy. It has two digital output bits as shown by the input-output characteristic of Fig. 2.8a, where possible digital outputs are 00, 10, and 01. The 00 output indicates that the sampled input is certainly negative, while the 10 output indicates that it is certainly positive. The 01 output indicates indecision, i.e., the quantizer does not know if the sampled input is either positive or negative [105]. This decision is postponed to subsequent stages and the correction logic (with the digital outputs of these stages) will correct this indecision. All this is only possible if the quantizer's offsets are less than $\pm 1/2 V_{LSB}$.

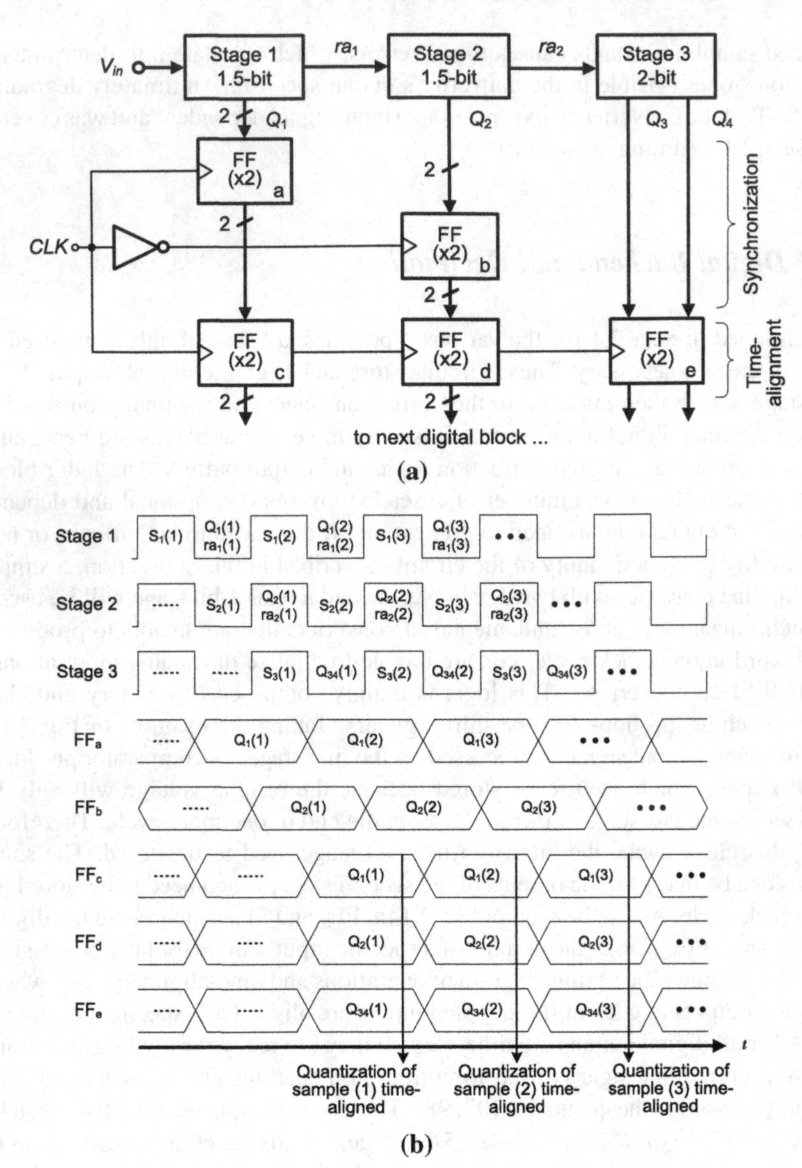

Fig. 2.14 Simple example of synchronization: **a** Pipeline converter, synchronization and time-alignment logic. **b** Time scheme of operations

An example of how correction logic operates is shown in Fig. 2.15. In Fig. 2.15a, an ideal situation is depicted, i.e., the comparators do not have offsets. In Fig. 2.15b, one of the comparators of stage 2 has a positive offset (indicated by the curly arrow). Even though there is an offset, the final output is the same as in the situation with no offset. This example shows how digital correction corrects for nonidealities in the

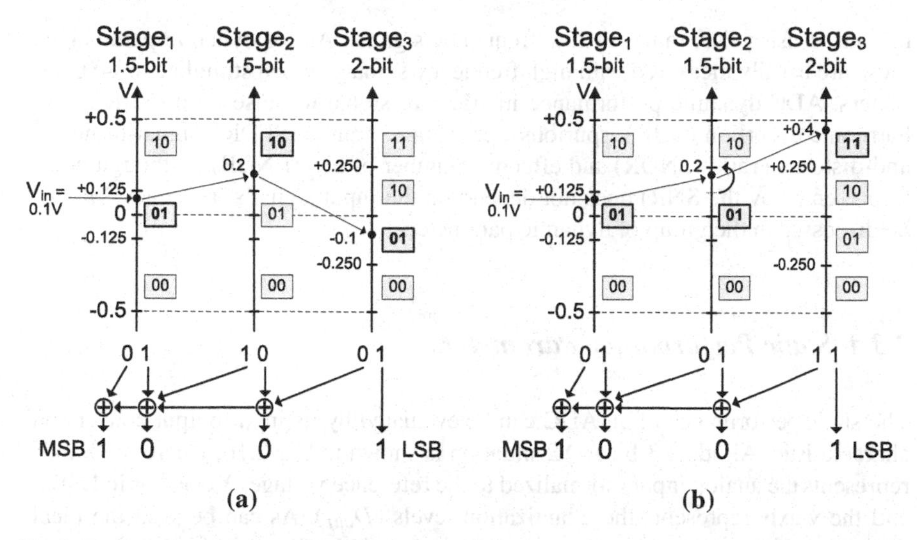

Fig. 2.15 Example of the operation of digital correction: **a** Ideal situation (and indecision corrected). **b** Offset error in stage 2: offset corrected

quantizers. Referring back to the example of Fig. 2.15a to demonstrate how digital correction also corrects indecisions. It is shown that the indecision of stage 1, i.e., stage 1 can not determine if the input is positive or negative (hence the 01 output), is corrected by the outputs of stages 2 and 3. This can be verified because the final result of the MSB (the sign bit of the digital output word) is 1, which indicates that the input signal is positive ($V_{in} = +0.1\,\mathrm{V}$).

In high-speed converters, the digital output buffers and digital output pads are one of the main contributors to the increase in substrate (ground) noise. The constant switching, charging, and discharging of the output nodes, couple undesired digital noise into the substrate which affects the normal operation of the analog circuits. In order to reduce the digital noise, the speed of operation (clocking frequency) needs to be reduced. So, instead of acquiring all the outputs (at the maximum clock frequency), if only one output in every N is acquired, no information is lost and a lower clock frequency (now divided by N) permits reducing the digital noise. This technique is called decimation by a factor of N, and is explained more thoroughly further on in the book.

2.3 Performance Metrics of Analog-to-Digital Converters

It will be the objective of this section to present metrics and parameters that characterize the performance of ADCs (in general). These are divided into two groups: static and dynamic parameters. Static parameters discussed here will be offset, gain, differential nonlinearity (DNL) and integral nonlinearity (INL). These are usually measured using DC, ramp, or low frequency signals. As for dynamic parameters, these are usually measured with high frequency signals, which stimulate these parameters. ADC dynamic performance metrics are signal-to-noise ratio (SNR), total harmonic distortion (THD), spurious free dynamic range (SFDR), signal-to-noise-and-distortion ratio (SNDR) and effective number of bits (ENOB). Although noise (represented by the SNR) does not depend on the input signal's frequency,[5] it will be discussed in the group of dynamic parameters.

2.3.1 Static Performance Parameters

The static performance of an ADC can be evaluated by its input-output conversion characteristic. An ideal 3-bit ADC situation is shown in Fig. 2.16, where the x-axis represents the analog input (normalized to the reference voltage, X_{REF},[6] or in LSBs) and the y-axis represents the quantization levels (D_{out}). As can be seen the ideal characteristic has a staircase waveform, where the width of each step is 1 LSB except for the first and last steps. The transition levels are taken in the middle of each analog input interval. The quantization (or amplitude discretisation) of an analog voltage, which intrinsically has infinite levels of quantization (zero error), by an ADC in eight quantization levels introduces an error. Graphically, this error is given by the difference between the staircase and the midpoint interpolating line (dashed line of the top graph of Fig. 2.16) and should ideally be limited between $\pm 1/2$ LSB as shown at the bottom of Fig. 2.16. This quantization error translates into an additive noise, i.e., the quantization noise. More on this noise is given in the dynamic performance parameters subsection.

With the understanding of Fig. 2.16, it is possible to describe most A/D converter static performance parameters.

2.3.1.1 Offset Error

Offset error is the horizontal difference between the first transition level of the real and the ideal ADC, as shown in Fig. 2.17a. It describes a shift for an analog input of 0 V, and graphically, the conversion characteristic is shifted horizontally. An expression for the offset error is given by

$$E_{offset} = \frac{x_{T_1} - x_{T_{1_{ideal}}}}{\Delta} \quad \text{(LSB)}, \tag{2.2}$$

[5] Intrinsic ADC noise is related with quantization and thermal noise. Other types of noise, such as, jitter and substrate noise (due to digital switching) are partially and often considered extrinsic to the A/D converter.

[6] X and x are used because they represent either a voltage or a current.

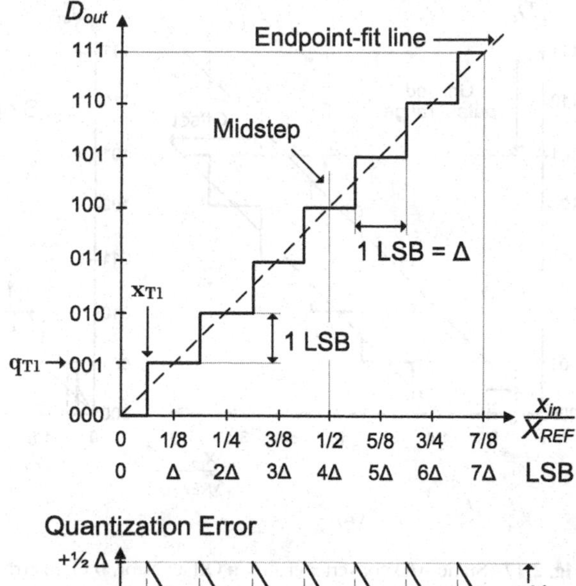

Fig. 2.16 Input–output conversion characteristic and quantization error of an ideal 3-bit A/D converter

where $\Delta = x_{LSB} = x_{REF}/2^N$ and N is the resolution of the ADC.

2.3.1.2 Gain Error

Gain error is the slope difference of the midpoint interpolating line of the real and ideal characteristics, as shown in Fig. 2.17b. For converter gains <1 and offset errors the output range of the ADC is limited. For gains > 1, a range of inputs have the same output (depicted in Fig. 2.17b). For the ideal case, this slope is unity. An expression for the gain error is given by

$$E_{gain} = \frac{q_{T_{2^N-1}} - q_{T_1}}{x_{T_{2^N-1}} - x_{T_1}} - 1 \quad \text{(LSB)}. \tag{2.3}$$

2.3.1.3 Differential Nonlinearity

In the ideal characteristic of Fig. 2.16, the horizontal difference between two consecutive transitions is exactly 1 LSB. In the characteristic of a real converter, any deviation from 1 LSB, between consecutive transitions, causes a differential

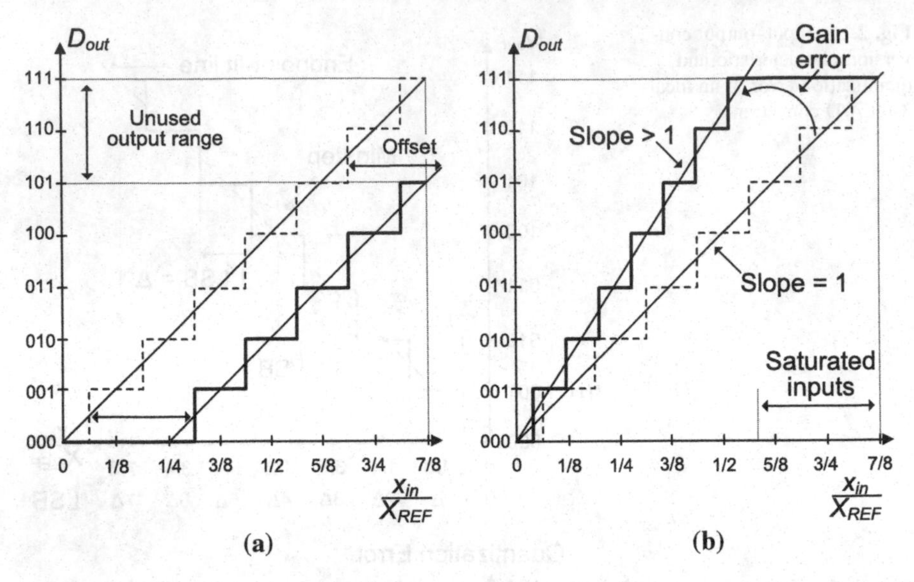

Fig. 2.17 Static A/D converter errors: **a** Offset error. **b** Gain error. Ideal characteristic represented by *dashed line*

nonlinearity (DNL) error. Figure 2.18 exemplifies some DNL errors. In an expression, the DNL can be given by,

$$\mathrm{DNL}(i) = \frac{x_{T_{i+1}} - x_{T_i}}{\Delta} - 1 \quad (\mathrm{LSB}), \quad i = 1, \ldots, 2^N - 2. \tag{2.4}$$

The DNL is usually taken as the max ($|\mathrm{DNL}(i)|$) for all i. Before determining the DNL for each code, offset and gain must be removed. This can be achieved using the endpoint-fit line, which creates a straight line from the first (origin) to the last (full-scale) code. The DNL profile is usually characterized by a graph with the digital output codes ($1 \ldots 2^N - 2$) for the x-axis and Eq. 2.4 for the y-axis. An example of this is shown at the bottom of Fig. 2.18a.

Observing Eq. 2.4, it is worth noting some special cases:

- $\mathrm{DNL}(i) = 0$: two consecutive transitions are equal to 1 LSB. Also true for the first and last codes, due to the endpoint-fit line system.
- $\mathrm{DNL}(i) = -1$: two consecutive transitions are equal, which means there is a missing quantization level, or missing code.
- $\mathrm{DNL}(i) \geq +1$: two consecutive transitions are larger than 1 LSB. High probability of the existence of missing codes in the DNL profile [9].

The DNL characteristic provides information about the converter's behaviour code by code. This means that the way in which each output code is encoded (the encoding process), has an effect on the converter's linearity [77]. The encoding process depends on the DAC architecture (capacitor matching) used in the MDAC

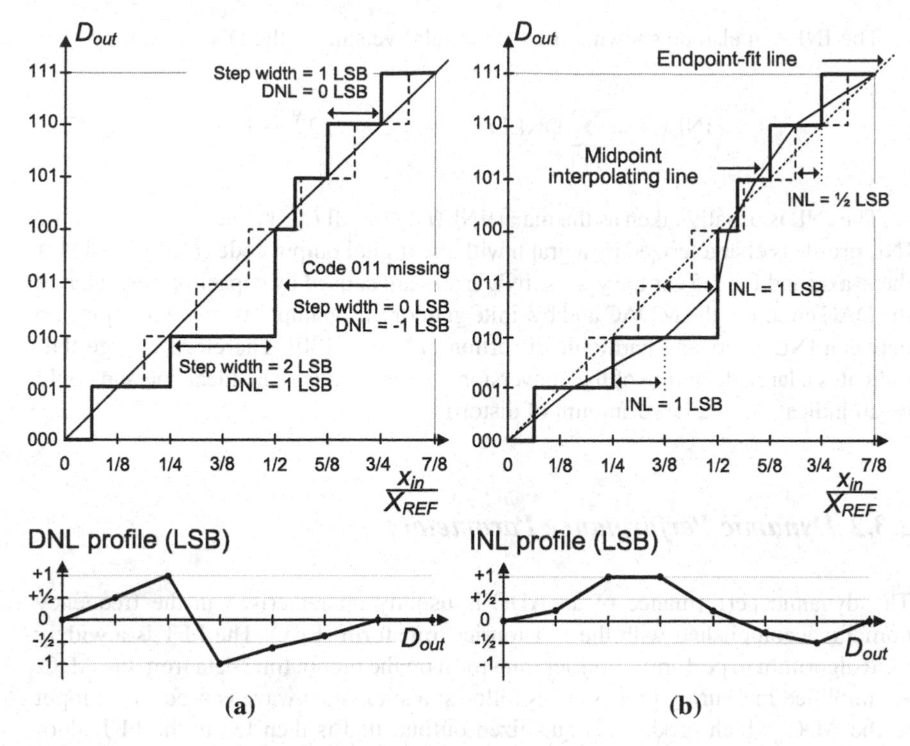

Fig. 2.18 Static A/D converter errors: **a** Differential nonlinearity (DNL). **b** Integral nonlinearity (INL). Ideal characteristic represented by the *dashed line*

circuit, and each architecture must be analysed separately [170]. Besides nonlinearity errors contributed by the MDAC circuits, the local stage quantizers also introduce errors with offsets. Fortunately, the latter can be corrected by the digital correction logic. Given that DNL errors cause the quantization characteristic to be different from the ideal one and quantization errors introduce additive noise, then DNL errors will also translate into an additive noise component called DNL noise, degrading the converter's SNR [77, 105].

2.3.1.4 Integral Nonlinearity

Integral Nonlinearity (INL) is a measure of the horizontal difference between each transition and its corresponding ideal transition. Using the endpoint-fit line to remove gain and offset errors, the INL becomes the difference from the midpoint interpolating line to the endpoint-fit line (straight line that connects first and last transition), as shown in Fig. 2.18b. INL can be defined by the following expression,

$$\mathrm{INL}(i) = \frac{x_{T_i} - \Delta(i-1) - x_{T_1}}{\Delta} \quad \text{(LSB)}, \quad i = 1, \ldots, 2^N - 1. \tag{2.5}$$

The INL can also be shown to be the cumulative sum of the DNL, given by,

$$\mathrm{INL}(i) = \sum_{j=1}^{i-1} \mathrm{DNL}(j), \quad i = 2, \ldots, 2^N - 1. \tag{2.6}$$

The INL is usually taken as the max ($|\mathrm{INL}(i)|$) for all i. Like the DNL profile, the INL profile is characterized by a graph with the digital output codes ($1 .. 2^N - 1$) for the x-axis and Eq. 2.5 for the y-axis. INL errors are caused by capacitor mismatch in the DAC circuit of the MDAC and by finite gain of the opamp. There is a relationship between INL errors and harmonic distortion [42, 105, 130]. Therefore, a large INL indicates a large deviation of the conversion characteristic to the ideal one and could be an indication of a large amount of distortion.

2.3.2 Dynamic Performance Parameters

The dynamic performance of an ADC is usually characterised in the frequency domain, accomplished with the fast fourier transform (FFT). The FFT is a widely used algorithm to perform a Fourier transform on the output time data from the ADC. A simplified measuring process is as follows: a sinusoidal wave is used as the input to the ADC, which produces a quantized output, that is then fed to the FFT algorithm and, finally, a spectrum of the quantized input signal is produced. The output spectrum permits measuring all the dynamic parameters that will be described here. To aid the enumeration and description of these parameters, a hypothetical FFT of the output of a two-channel time-interleaved ADC (some errors exaggerated) shown in Fig. 2.19 will be used. For the equations presented below, the band of interest is considered the Nyquist bandwidth ($F_{S/2}$).

2.3.2.1 Signal-to-Noise Ratio

As the name indicates it is the ratio of the signal power to the noise power. Observing Fig. 2.19, the noise power excludes DC, signal, and harmonic components, but includes the spurious tones due to time-interleaving mismatches. The theoretical maximum signal-to-noise ratio (SNR) that an ADC can achieve is (only taking into account quantization noise),

$$\mathrm{SNR} = 6.02N + 1.76 \quad (\mathrm{dB}), \tag{2.7}$$

where N is the resolution of the converter. As already stated throughout this chapter, there are other noise sources which contribute to the total noise of the ADC (further degrading its SNR), which are clock jitter, DNL errors, and thermal noise. A more complete expression becomes,

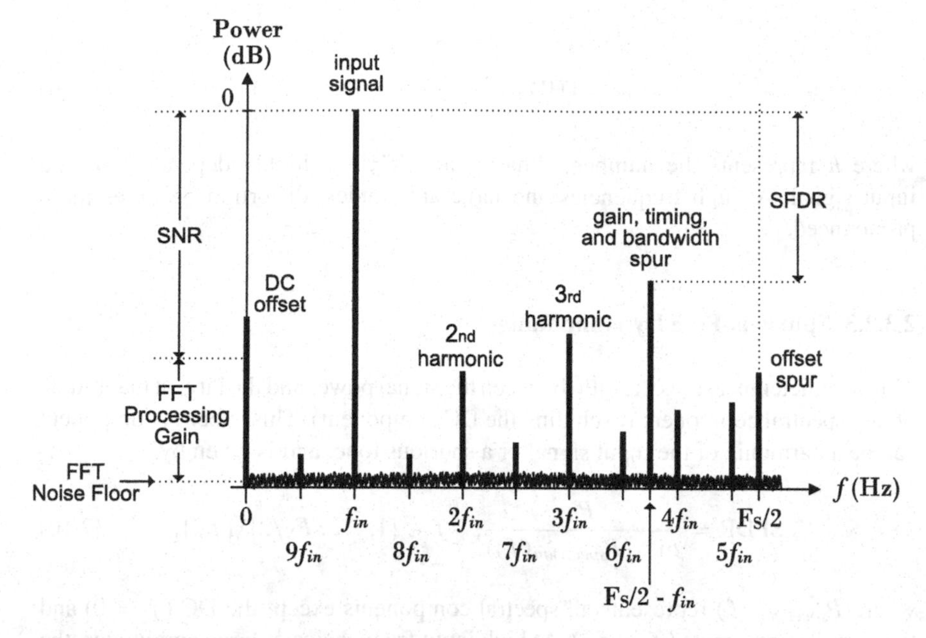

Fig. 2.19 Example of an FFT of a two-channel time-interleaved ADC

$$\text{SNR} = \frac{P_S}{P_j + P_{q+DNL} + P_{th}}, \tag{2.8}$$

where P_j, P_{q+DNL}, and P_{th} are the jitter, quantization plus DNL, and thermal noise power, respectively. Therefore, the overall SNR is input signal dependent (due to extrinsic effects): high signal frequency increases jitter and small signal amplitude reduces signal power. To obtain the result in decibels (dB), 10 log of the respective equation should be taken.

As shown in Fig. 2.19, the noise floor is situated such that each noise bin (discrete line) is, on average, below the full-scale by $6.02N+1.76$ plus the FFT processing gain (given by $10\log(n_{points}/2)$, where n_{points} is the number of points used to compute the FFT). This processing gain is particularly helpful when trying to distinguish harmonics from noise. The more points in the FFT, the lower the noise floor will be, but the harmonics will stay at their original magnitude.

2.3.2.2 Total Harmonic Distortion

When an input is quantized by a nonideal ADC, tones appear, in the output spectrum, at multiples of the signal's frequency. These tones are called harmonics, and the total harmonic distortion (THD) measures the ratio of the sum of the harmonics' power, P_H, to the signal's power (P_S). In an expression this is given by

$$\mathrm{THD} = \frac{\sum_{i=2}^{h} P_H(i)}{P_S}, \tag{2.9}$$

where h represents the number of harmonics. THD is highly dependent on the input signal. At high frequencies and large amplitudes, distortion becomes more pronounced.

2.3.2.3 Spurious-Free Dynamic Range

This parameter measures the ratio between the signal power and the largest magnitude of any spectral component (excluding the DC component). This spectral component can be a harmonic of the input signal or a spurious tone, and is given by,

$$\mathrm{SFDR} = \frac{P_S}{\max(P_{spectrum}(f))}, \quad f \in \{1, \ldots, F_S/2\} \backslash \{f_{in}\}, \tag{2.10}$$

where $P_{spectrum}(f)$ represents all spectral components except the DC ($f = 0$) and the signal components ($f = f_{in}$). At high input frequencies or large amplitudes, the limiting tone will probably be a harmonic, whereas at low amplitudes a spurious tone could limit the SFDR.

2.3.2.4 Signal-to-Noise-and-Distortion Ratio

This parameter is a complete indication of the overall dynamic performance of the converter. It is complete in the sense that it combines all performance degradation elements, such as, the noise sources of the SNR and the distortion components of the THD. In an expression the SNDR can be given by

$$\mathrm{SNDR} = \frac{P_S}{\sum_{i=2}^{h} P_H(i) + P_j + P_{q+DNL} + P_{th}} = 10^{\frac{\mathrm{THD}}{10}} + 10^{\frac{-\mathrm{SNR}}{10}}. \tag{2.11}$$

2.3.2.5 Effective Number of Bits

Given that the SNDR of Eq. 2.11 will be less than the theoretical SNR limit given by Eq. 2.7, it becomes important to define a real resolution for the converter. This real resolution is known as the effective number of bits (ENOB) and an expression for it can be obtained by solving Eq. 2.7 for N. This is given by

$$\mathrm{ENOB} = \frac{\mathrm{SNDR} - 1.76}{6.02} \text{ bits}. \tag{2.12}$$

2.4 Overview and Comparison of Published Work

This section presents an overview of the state-of-the-art concerning the work carried out in this book. Given that this work presents results from silicon prototypes of two different circuits, namely an opamp and an ADC, an overview of published work related with these circuits is presented.

Besides the overviews of these two circuits, a review of published data concerning ADC reference voltage circuits will also be given. The objective of this review is to obtain some insight and criteria about the power dissipated and area occupied by these circuits in the context of A/D conversion. This review is fundamental because the ADC described in this book precludes all reference voltage circuits, therefore saving power and area. To be able to quantify, in average, the power and area saved by the proposed techniques, the results from this review will become useful.

2.4.1 Two-Stage Opamps

The following overview, summarized in Table 2.3, concerns class-A and class-AB two-stage opamps from the past ten years, simulated or fabricated in a CMOS technology with GBW > 30 MHz. In order to evaluate and compare the performance of the various opamps, a figure-of-merit (FoM) will be used [118]. This FoM basically evaluates the speed to current consumption ratio. The better the opamp, the smaller the FoM. However, this becomes a problem because the values of the FoM become too small, which in turn, makes it difficult to comprehend and compare, so the inverse of the FoM of [118] will be used instead, which is given by

$$\mathrm{FoM_{OA}} = \frac{\mathrm{GBW} \cdot C_L}{\mathrm{Power}} \quad [\mathrm{MHz} \cdot \mathrm{pF/mW}]. \tag{2.13}$$

Regarding Table 2.3, besides the technology and the supply voltage used, various AC and transient response performance parameters are given. The settling time and settling error are represented by T_S and $T_{S\varepsilon}$, respectively. Figure 2.20 depicts the $\mathrm{FoM_{OA}}$ of the opamps of Table 2.3 versus their respective GBW. A linear interpolation line (dashed line) in Fig. 2.20 demonstrates the average value of $\mathrm{FoM_{OA}}$(GBW). The negative slope of this line, although small, demonstrates the difficulty in achieving high GBW with high energy efficiency. This challenge is well observed by the marked ([a], [b], and [c]) opamps. Opamp [b] [149] represents a very high GBW opamp with a given $\mathrm{FoM_{OA}}$, while opamp [a] [73] achieves more than two times the $\mathrm{FoM_{OA}}$ but at the cost of a much lower GBW. The most energy efficient opamp ([c]) is reported in [73], and achieves a $\mathrm{FoM_{OA}}$ of 1213 MHz·pF/mW by employing a dynamic threshold technique and controlling the bulk voltage of critical transistors.

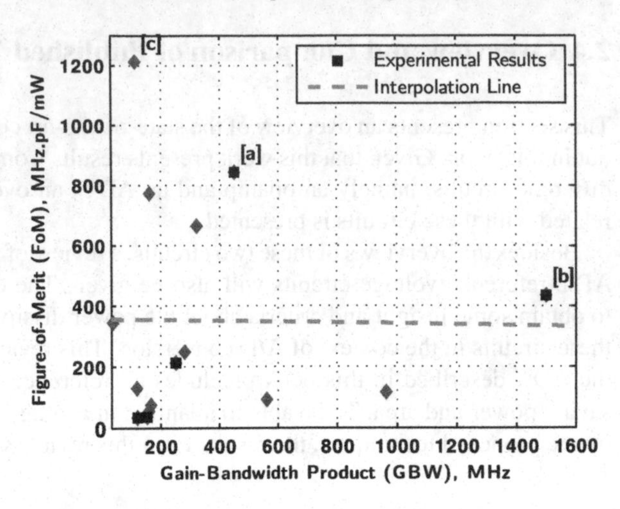

Fig. 2.20 State-of-the-art two-stage opamps with GBW > 30 MHz designed in CMOS technologies

Table 2.3 Overview of two-stage opamps from the past ten years with GBW > 30 MHz in CMOS technologies

Ref.	Tech. [μm]	V_{DD} [V]	DC Gain [dB]	$T_{S\varepsilon}$ [%]	T_S [ns@V_{pp}]	C_L [pF]	Power [mW]	GBW [MHz]	FoM_{OA} $\left[\frac{MHz \cdot pF}{mW}\right]$
[73]	0.13	0.5	51.0	N/A	N/A	6.5	0.6	112	**1213**
[159][a]	0.065	1.0	56.1	1	8@0.5	3	1.6	450	**844**
[132]	0.18	1.8	74	N/A	N/A	1.75	0.362	160	**773**
[147]	0.13	1.2	77	0.1	3.8@0.1	4	1.94	322	**664**
[149][a]	0.12	1.2	40.4	0.1	9.2@0.1	3.2	11	1500	**436**
[2]	0.18	1.0	64.4	N/A	N/A	5	0.522	36	**345**
[160]	0.25	1.5	78	0.24	7.2@2	8	9	280	**249**
[162][a]	0.5	2.5	90	1	15@0.7	12	14	250	**214**
[184]	0.25	1.5	69.5	0.01	20.1@0.5	6	5.5	119	**130**
[182]	0.25	2.5	72	1	2.1@0.6	3	25	963	**116**
[152]	0.5	1.8	75	N/A	N/A	1.8	11	562	**92**
[183]	0.25	1.5	80	0.01	7.1@0.5	4	8.9	167	**75**
[143]	0.13	1.2	68.7	N/A	N/A	1.57	4.51	160	**56**
[179][a]	0.5	3.0	84	0.1	47@1.2	10	43	153	**36**
[72][a]	0.5	3.5	90	0.1	10@1	1	3.5	120	**34**

[a]Experimental results

2.4.2 *Medium-Low Resolution High-Speed MDAC-Based ADCs*

The following state-of-the-art concerns medium-low resolution high-speed MDAC-based ADCs. The resolutions chosen are 6–8 bit, with ENOB > 5 bits and sampling frequencies, F_S > 200 MS/s. Only the MDAC-based architectures are chosen for a more fair comparison with the prototyped ADC described, as well as, the proposed techniques. In order to evaluate and compare the performance of the various ADCs, it

Table 2.4 Overview of 6–8 bit MDAC-based ADCs with ENOB > 5 bits and F_S > 200 MS/s designed in a CMOS technology

Ref.	Tech. [μm]	Resolution [bits]	F_S [MS/s]	ENOB [bits]	Power [mW]	Area [mm^2]	FoM_{ADC} $\left[\frac{fJ}{conv.\text{-}step}\right]$
[79]	0.09	8	320	7.3	12.8	0.53	**253**
[168]	0.065	8	800	7	30	0.12	**283**
[15]	0.18	8	200	6.4	8.5	0.05	**503**
[90]	0.18	8	200	7.7	30	0.15	**731**
[83]	0.18	8	200	7	22	0.32	**830**
[26]	0.13	6	1000	5.3	49	0.16	**1240**
[75]	0.09	7	550	5.7	60	0.37	**2045**
[75]	0.09	7	1100	5.7	92	0.37	**2206**
[134]	0.09	8	250	6.2	22.8	0.81	**2580**
[115]	0.09	6	10300	5.1	1600	N/A	**4699**
[136]	0.35	8	4000	6.1	4600	28.85	**33531**

is necessary to use an appropriate FoM. Although highly contested, the FoM mostly used and found in the literature is the Walden FoM [173] (or its inverted form), given by

$$FoM_{ADC} = \frac{\text{Power}}{2^{\text{ENOB}} \cdot \min\{F_S, 2BW\}} \quad [\text{J/conv.} - \text{step}], \tag{2.14}$$

where $\min\{F_S, 2BW\}$ represents the minimum between the sampling frequency and two times the ADC's bandwidth (BW). This FoM weighs the ADCs overall performance given by its speed (F_S) and linearity (ENOB, which includes SNR and THD), to its total power consumption. Unlike the FoM_{OA}, the smaller the FoM_{ADC}, the more energy efficient an ADC is.

Table 2.4 presents the state-of-the-art of the aforementioned medium-low resolution high-speed MDAC-based ADCs. Besides showing ENOB, F_S, power consumption, and FoM_{ADC}, the table also shows each ADC's resolution and occupied area, as well as the employed technology. The table only contemplates ADCs with experimental results.

Figure 2.21 shows the results of the FoM of the ADCs of Table 2.4 plotted against their respective sampling frequency (Fig. 2.21a) and ENOB (Fig. 2.21b). The latter, plots the FoM against normalized ENOB because the state-of-the-art consists of ADCs with different resolutions. Therefore, the various ADCs can be compared in terms of their linearity. The interpolation line of Fig. 2.21a clearly demonstrates the difficulty in achieving a good FoM at high sampling frequencies.

In Fig. 2.21 the ADC marked as [a] [79] represents the converter with the best FoM, while [b] [115]) marks the one with the highest sampling frequency. These two ADCs exemplify that for increasing sampling frequencies, the FoM degrades. Note however that, they were designed with different objectives. Concerning ADC [a], it achieves its FoM by using a multilevel power optimization algorithm based on geometric

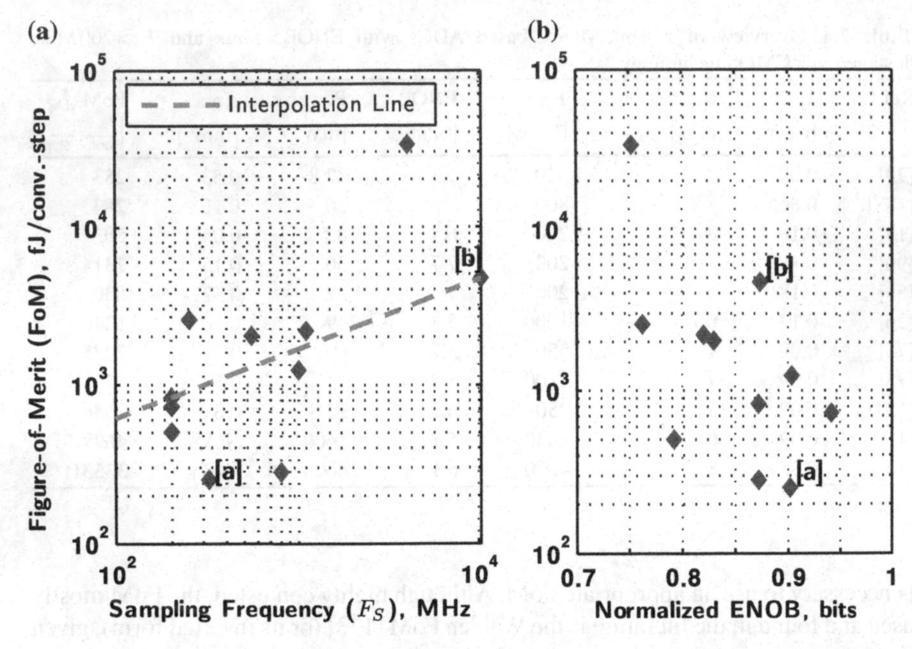

Fig. 2.21 The state-of-the-art of 6–8 bit MDAC-based ADCs with ENOB > 5 bits and F_S > 200 MS/s in CMOS technologies: **a** FoM versus F_S. **b** FoM versus normalized ENOB

programming. However, it needs two supply voltages, 1.2 and 2.1 V for digital and analog circuits, respectively. The total power of the ADC did not contemplate the power of the reference voltage buffers. ADC [b] uses digital calibration to correct for the nonidealities that arise from using open-loop amplifiers and mismatches between the interleaved channels. Besides calibration, each comparator uses a trimming circuit and each unit ADC is composed of two separate ADCs, one will be in operation while the other will be in calibration. Regarding calibration, only the ADCs with a high number of interleaved channels, namely, [115, 136] use digital calibration. What concerns reference voltage circuit power, only [79] indicates this power (but does not include it in the overall ADC power), while all others have omitted this value or have not indicated if it is included in the total power. Moreover, adding the reference power indicated in [79] to its total ADC power, degrades its FoM from 253 to 365 fJ/conv.-step.

2.4.3 ADC Reference Voltage Circuitry

The objective of this subsection is to obtain some knowledge and criteria concerning reference voltage circuitry related to ADCs. Reference voltage circuitry can be understood as the circuits used to generate, buffer, and decouple a reference voltage. As already mentioned, reference voltage buffers are one of the most power

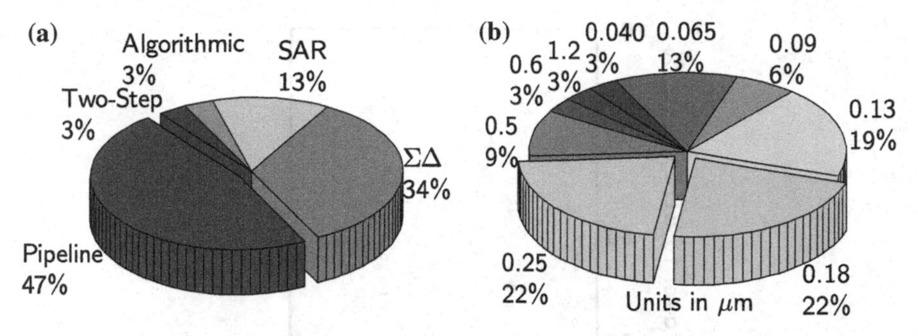

Fig. 2.22 Overview of the ADC reference voltage circuitry data set's characteristics: **a** Architectures. **b** Technology nodes

consuming blocks of ADCs. Besides consuming power, these circuits occupy a huge amount of silicon area due to the necessary decoupling capacitors (and damping resistors). In some cases these capacitors need to be so large and impractical to implement on-chip, they are placed off-chip, which brings other problems (explained in Sect. 2.2.5).

It is not an easy task compiling this overview because most articles found in the literature do not present the power consumption and area occupied by the reference voltage circuits. Nevertheless, the articles obtained that discriminate the power and area (of the various blocks that compose an ADC) will be enough to obtain some insight into this issue.

For this overview the experimental data from these references is used [4, 6, 8, 12–14, 16, 19, 20, 23, 24, 31–34, 36, 40, 53, 54, 70, 78, 81, 117, 125, 126, 129, 155, 157, 166, 168, 169, 187]. Figure 2.22 is used to characterize the data set for a better comprehension of the ADCs and their characteristics. Figure 2.22 depicts the architectures and the percentage of each architecture in the total population of the data set, while Fig. 2.22 shows the technologies used to implement the ADCs and their percentage of the total population. As mentioned before, the main objective of this overview is to obtain a rough estimate of the power and area of the reference voltage circuits compared to the ADC's overall power and area, respectively. From the mentioned data set, the percentage of the power consumed by reference voltage circuits (to the overall ADC power) is, in average, 29 %, while the area occupied by these circuits (to the overall ADC area) is, in average, 19 %. This area excludes off-chip decoupling capacitors.

From the data set it is possible to extract other interesting information, namely, the relationship between the reference circuit's power and: (a) its area, (b) the sampling frequency, and (c) the resolution of the ADC. These relationships are depicted in Fig. 2.23. The interpolation line of Fig. 2.23a clearly indicates a direct relationship between the power and area of the reference circuit. The interpolation line of Fig. 2.23b shows a smaller direct correlation with F_S, while Fig. 2.23c shows an even smaller relationship with the ADC's resolution.

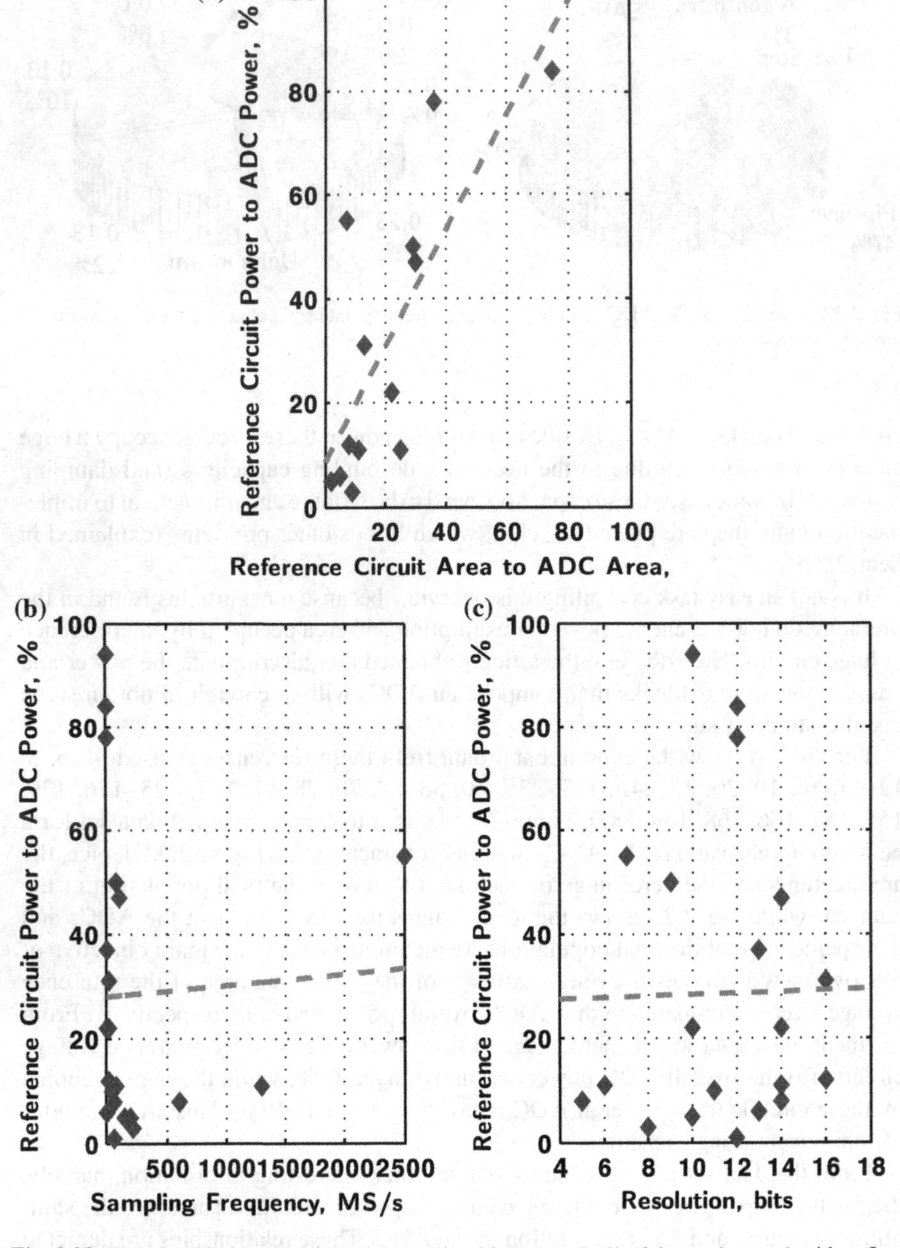

Fig. 2.23 Relationship between the reference circuit's power (in % of the total power) with: **a** Its area. **b** ADC's sampling frequency. **c** ADC's resolution. *Dashed line* represents the interpolation of the data points

Given such a small data set, it is important to note and remember that all these estimates and graphs correspond to a very rough representation regarding the power and area of reference voltage circuits in the context of A/D conversion.

This concludes the overview and comparison with published work. The following chapter presents the proposed MDAC circuits.

Chapter 3
Capacitor Mismatch-Insensitive Multiplying-DAC Topologies with Unity Feedback Factor

Abstract This chapter describes two capacitor mismatch-insensitive MDAC architectures with unity feedback factor. First, the conventional MDAC will be described and analysed, to be compared with the proposed circuits. Then, each one of the proposed MDACs will be discussed. Besides describing the basic concept, various analyses are carried out, namely, gain error, reference shifting error, feedback factor, and noise. These analyses show the benefits and limitations of each MDAC circuit. To conclude the section a comparative analysis is carried out with other MDAC and multiply-by-two amplifiers (MBTAs).

3.1 Conventional MDAC

3.1.1 Principle of Operation

The conventional and well known 1.5-bit MDAC, initially proposed in [98], is shown in Fig. 3.1. This circuit operates in two phases. During ϕ_1, the input signal is sampled on capacitors C_{11} and C_{21} for the positive signal path, and on C_{12} and C_{22} for the negative signal path. During the residue amplification phase, ϕ_2, C_{11} and C_{12} are inserted in the feedback loop and the charge stored on C_{21} and C_{22} is redistributed with the former capacitors. For equally sized capacitors, the resulting differential output voltage at the end of ϕ_2, will be two times the differential input voltage sampled in ϕ_1. In addition, the output voltage will be affected by a level shifting equal to $\pm V_{REF}$ depending on the MDAC's operation mode (X or Z). The X, Y, and Z operation modes are defined by the local 1.5-bit quantizer. In Fig. 3.1, $X * 2$ (with $X = 1$ as an example) indicates that the respective switches only close if the MDAC is in X operation mode and the current phase is ϕ_2.

As a result, the ideal 1.5-bit MDAC characteristic, or transfer function (TF), is given by

$$V_{od} = 2V_{id} + B \cdot V_{REF}, \tag{3.1}$$

M. Figueiredo et al., *Reference-Free CMOS Pipeline Analog-to-Digital Converters*, Analog Circuits and Signal Processing, DOI: 10.1007/978-1-4614-3467-2_3,

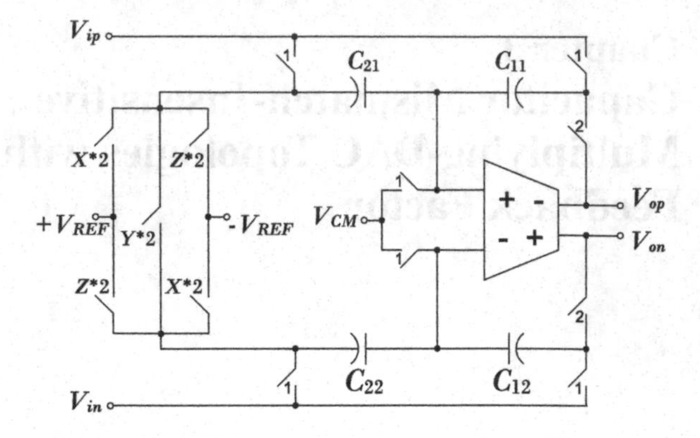

Fig. 3.1 Conventional fully differential 1.5-bit MDAC

where $V_{od} = V_{op} - V_{on}$, $V_{id} = V_{ip} - V_{in}$, V_{REF} is the MDAC's reference voltage, and $B = +1$ (if $X = 1$), 0 (if $Y = 1$), -1 (if $Z = 1$). Through this characteristic two sources of error may be found. The first is related with the gain term affecting V_{id}, which should ideally be 2, and the other with the level shifting terms of V_{REF},[1] which should be +1, 0, or -1.

3.1.2 Gain and Reference Shifting Error Analysis

For all MDAC analyses carried out in this chapter, unless otherwise indicated, an ideal opamp and ideal switches will be considered. Furthermore, whenever possible, the single-ended versions of the MDACs will be analyzed (for simplicity reasons), but the final equations and graphs of each analysis will always correspond to the fully differential version of the MDACs.

The exact TF of the conventional MDAC in the presence of parasitic capacitors is derived next. The single-ended version of this MDAC is shown in Fig. 3.2. Considering its configuration during ϕ_2 (Fig. 3.2c), its TF is defined solely by the charge conservation equation derived at the inverting input of the opamp, i.e.,

$$-V_i C_{11} - V_i C_{21} = -V_o C_{11} + V_{REF} C_{21}, \tag{3.2}$$

where it is assumed $X = 1$. By solving Eq. 3.2 for the output voltage, V_o, results

$$V_o = \left(1 + \frac{C_{21}}{C_{11}}\right) V_i + \frac{C_{21}}{C_{11}} V_{REF}. \tag{3.3}$$

[1] The error related with V_{REF} is of minor importance, as long as this voltage is provided by a bandgap circuit (i.e., stable over time), the MDAC's residue characteristic maintains its symmetry and the reference shifting errors are the same for all pipelined stages.

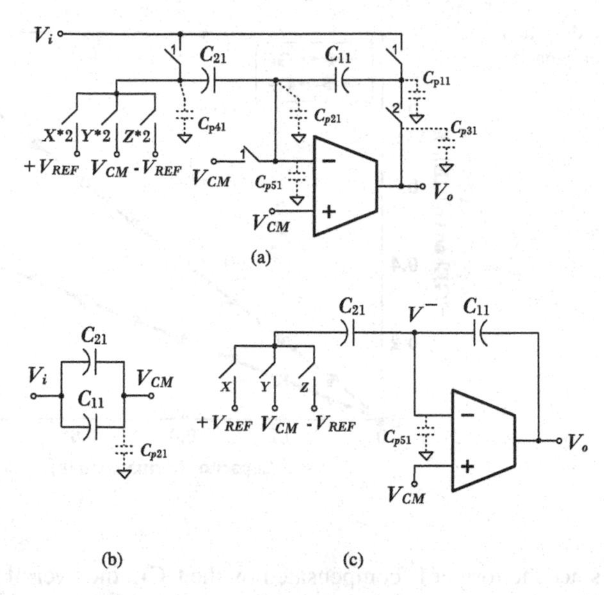

Fig. 3.2 **a** Conventional single-ended 1.5-bit MDAC with parasitic capacitors. MDAC configuration during **b** ϕ_1 and **c** ϕ_2

Equation 3.3 shows that the conventional MDAC is insensitive to parasitic capacitances. It also shows that if the capacitors have equal values, i.e., $C_{11} = C_{21} = C$, the ideal TF of Eq. 3.1 would be obtained. However, given that in current standard CMOS fabrication processes, the ratio accuracy of two capacitances is bounded to about approximately 0.2 %, which results in a gain error of about 0.1 % (or an equivalent resolution of 10 bits), the conventional MDAC has similar accuracy limitations. In other words, the accuracy of the conventional MDAC, and therefore ADCs that employ it, are limited to about 10 bits.

Extending this analysis to the fully differential MDAC implementation (Fig. 3.1) results in,

$$V_{od} = 2\left(\frac{1}{2} + \frac{C_{21}}{4C_{11}} + \frac{C_{22}}{4C_{12}}\right) V_{id} + \left(\frac{C_{21}}{2C_{11}} + \frac{C_{22}}{2C_{12}}\right) B \cdot V_{REF}, \tag{3.4}$$

The gain error (GE) is defined as the relative deviation of the term multiplying V_{id} in Eq. 3.4 from the desired value of 2. Likewise, the reference shifting error (RE) is the relative deviation of the term multiplying V_{REF} from the desired value of B.

Assuming $C_{ij} = C(1 + \varepsilon_{ij})$, with $i, j = \{1, 2\}$, where ε_{ij} are uncorrelated Gaussian random variables of the relative mismatch errors with zero mean and standard deviation σ, the GE and RE can be evaluated through high level Monte Carlo (MC) simulations using MATLAB [108]. Figure 3.3 depicts the σ[GE] and σ[RE] for different values of relative capacitor mismatch errors. Each relative mismatch data point is the result of 1000 MC runs. It can be seen that the RE is twice the

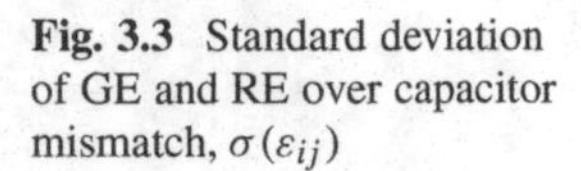
Fig. 3.3 Standard deviation of GE and RE over capacitor mismatch, $\sigma(\varepsilon_{ij})$

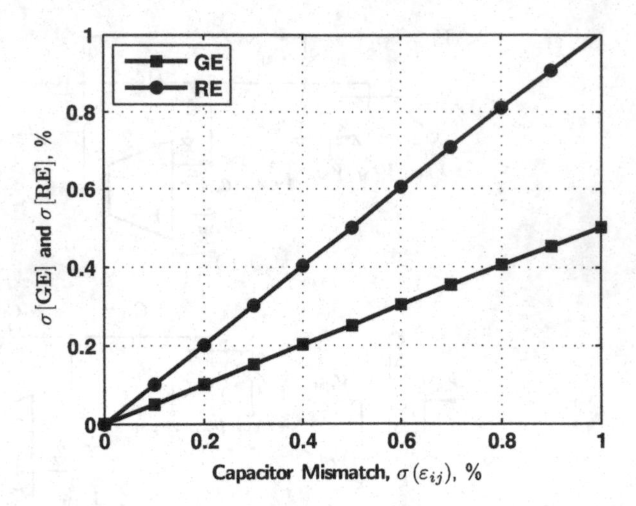

GE. However, since the former is compensated by the DCL, the overall accuracy of the MDAC is dictated by the gain error. At 0.2 % relative mismatch, the $\sigma[\text{GE}] \approx 0.1\,\%$, hence the 10-bit level limitation.

3.1.3 Feedback Factor

Along with the gain-bandwidth product (GBW) of the opamp, the dynamics of the MDAC are directly influenced by the feedback factor, β. Greater β means faster MDAC settling dynamics (for a given power budget). For the conventional MDAC, the β can be calculated with help of Fig. 3.2c (for $Y = 1$). Considering only C_{p51} as the dominant parasitic capacitance, i.e. the opamp's input capacitance, the β is expressed as

$$\beta = \frac{V^{-}(s)}{V_o(s)} = \frac{C_{11}}{C_{11} + C_{21} + C_{p51}}, \tag{3.5}$$

where $s = j\omega$. Consequently, for small parasitic capacitance ($C_{p51} \ll C_{11} + C_{21}$), the β approaches its maximum of 1/2 [98].

3.1.4 Thermal Noise Analysis

The lower bound of the MDAC's dynamic range is primarily imposed by thermal noise. This noise appears at the MDAC's output as a consequence of the thermal noise of the switches of both phases of operation and the thermal noise associated with the opamp. For the noise analysis shown in this chapter, it will be assumed, for simplicity,

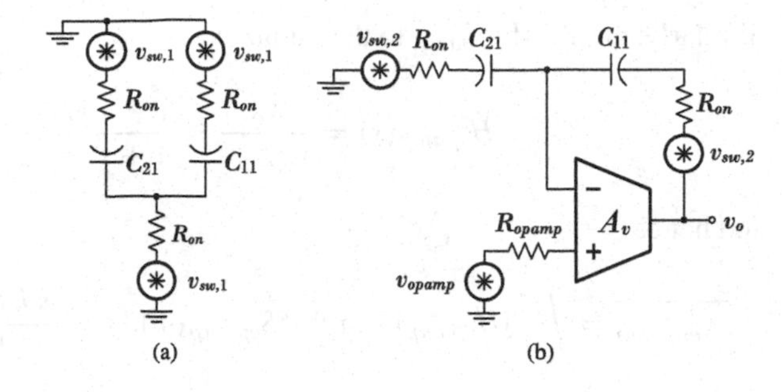

Fig. 3.4 Equivalent circuit for thermal noise analysis of the conventional MDAC in **(a)** ϕ_1 and **(b)** ϕ_2

that the switches are equally sized with noise voltage sources having an equal one-sided power spectral density (PSD) of $S_{SW}(f) = 4kTR_{on}$, where k is Boltzmann's constant, T is absolute temperature, and R_{on} represents the switches' ON-resistance. Furthermore, all noise sources are treated as uncorrelated, it is assumed that the capacitors are equally sized, i.e., $C_{11} = C_{21} = C$, and the opamp is modeled by a single pole transfer function, $A_v(s) = A_0/(1 + s/p_1)$, where A_0 and p_1 are the opamp's DC gain and dominant pole, respectively. For simplicity, the accumulated noise on the parasitic capacitor at the opamp's input is neglected. The noise analysis for ϕ_1 is performed using the equivalent circuit shown in Fig. 3.4a. Given the voltage noise source, $v_{SW,1} = 4kTR_{on}$, it is straightforward to show that the mean-square (MS) thermal noise voltage sampled on capacitors C_{11} and C_{21} at the end of ϕ_1 is kT/C [150]. The equivalent MS charge is kTC. During ϕ_2, the charge stored on C_{21} is completely transferred to C_{11}. As a result, the MS output voltage is given by

$$\overline{v_{osw,1}^2} = \frac{kTC_{11} + kTC_{21}}{C_{11}^2} = \frac{2kTC}{C^2} = \frac{2kT}{C}. \tag{3.6}$$

However, as shown in Fig. 3.4b, other thermal noise sources, namely two switches and the opamp, also contribute to the total output noise. The voltage noise source, v_{opamp}, represents the input-referred thermal noise of the opamp and has a PSD of $S_{opamp}(f) = 4kTR_{opamp}$, where R_{opamp} is the opamp's single-ended equivalent input-referred noise resistance, which is dependent on opamp topology and sizing.

The transfer functions from the voltage noise sources, $v_{SW,2}$, to the output can be approximated by

$$H_{SW,2}(s) = \frac{v_o}{v_{SW,2}} \approx \frac{C_{21}}{C_{11}} \frac{1}{1 + s/(\beta A_0 p_1)} = \frac{1}{1 + s/(\beta A_0 p_1)}, \tag{3.7}$$

assuming that $\beta A_0 \gg 1$. Note that the C_{21}/C_{11} term is only valid for the input switch. The MS output thermal noise voltage produced by both switches in ϕ_2 is given by

$$\overline{v_{osw,2}^2} = 2 \times \int_0^{\infty} |H_{SW,2}(j2\pi f)|^2 S_{SW,2}(f) \mathrm{d}f = \frac{2kTR_{on}}{1/(\beta A_0 p_1)}. \tag{3.8}$$

Similarly, the TF of v_{opamp} to the output is

$$H_{opamp}(s) = \frac{v_o}{v_{opamp}} \approx \frac{1/\beta}{1 + s/(\beta A_0 p_1)}, \tag{3.9}$$

and hence

$$\overline{v_{oopamp}^2} = \int_0^\infty |H_{opamp}(j2\pi f)|^2 S_{opamp}(f)\mathrm{d}f = \frac{kT R_{opamp}(1/\beta)^2}{1/(\beta A_0 p_1)}. \tag{3.10}$$

Therefore, the total output MS thermal noise is

$$\overline{v_o^2} = \overline{v_{osw,1}^2} + \overline{v_{osw,2}^2} + \overline{v_{opamp}^2} = kT\left[\frac{2}{C} + \beta A_0 p_1 \left(2R_{on} + \frac{R_{opamp}}{\beta^2}\right)\right]. \tag{3.11}$$

For the fully differential conventional MDAC implementation, the total MS noise can be written as

$$\overline{v_{od}^2} = 2 \times kT\left[\frac{2}{C} + \beta A_0 p_1 \left(2R_{on} + \frac{R_{opamp}}{\beta^2}\right)\right]. \tag{3.12}$$

The noise analysis of the conventional MDAC did not contemplate the noise contribution of the reference voltage (V_{REF}) circuitry. As already mentioned in the previous chapter, if these circuits are not adequately sized and decoupled, they will add noise to the MDAC's output, and can even be the ADC's limiting factor [113].

3.2 Current-Mode Reference Shifting MDAC

This section proposes an MDAC circuit with unity feedback factor that is insensitive to capacitor mismatch. Reference shifting occurs in current-mode during the amplification phase. As an abbreviation this MDAC will be called CMRS-MDAC.

3.2.1 Principle of Operation

The proposed unity β, capacitor mismatch-insensitive 1.5-bit MDAC is shown in Fig. 3.5a. Similarly to the conventional MDAC, this circuit operates in two clock phases. During ϕ_1, the differential input voltage is sampled on capacitors C_{1j}, C_{2j}, and C_{3j}, $j = \{1, 2\}$. During ϕ_2, capacitors C_{1j} and C_{2j} are associated in series around the opamp's feedback loop and the gain of two is obtained by voltage sum, instead of charge redistribution, as occurs in the conventional implementation. Reference shifting occurs (during ϕ_2) when current sources, I_P and I_N, are turned on. These

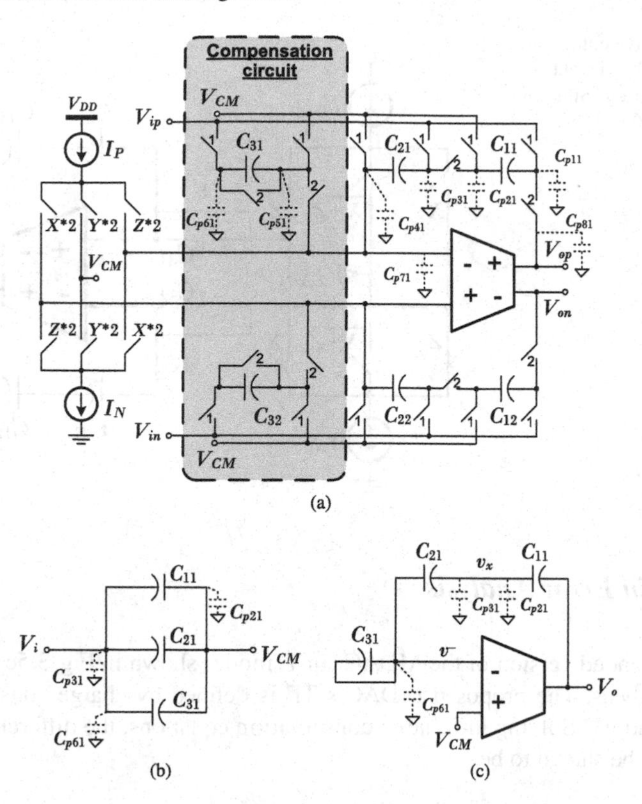

Fig. 3.5 **a** Fully differential enhanced feedback factor 1.5-bit mismatch-insensitive MDAC with current-mode reference shifting. **b** Single-ended equivalent circuit during sampling (ϕ_1). **c** Single-ended equivalent circuit during ϕ_2 for GE analysis

current sources sink/source current through the series associated capacitors changing the output voltage by an amount proportional to the respective current, feedback capacitance and duration of ϕ_2. By the end of ϕ_2 the output voltage should have changed by an amount equal to $\pm V_{REF}$ (differentially), for X and Z modes. This situation is illustrated in Fig. 3.6, where the dashed lines indicate the flow of current. Regarding the output waveforms, in Y operation mode it is exponential, but for X and Z it has a ramped integrating characteristic until the end of ϕ_2. Following this approach for obtaining a capacitor mismatch-insensitive gain of two, the circuit becomes sensitive to parasitic capacitors C_{p2j} and C_{p3j} (nodes between C_{2j} and C_{1j} in Fig. 3.5). To attenuate this sensitivity, capacitors C_{3j} are employed. Notice, however, that these capacitors are shorted during ϕ_2 because only the charge stored in their parasitic capacitors (C_{p6j}) is used to compensate the charge stored on C_{p2j} and C_{p3j}, as shown next.

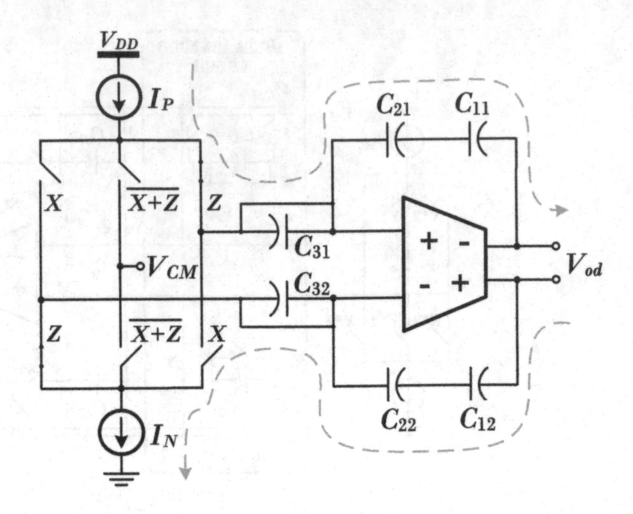

Fig. 3.6 Equivalent circuit to demonstrate the current-mode reference shifting. Situation depicted for $Z = 1$

3.2.2 Gain Error Analysis

The single-ended version of the MDAC (in Y mode) shown in Fig. 3.5c is used for the GE analysis. The proposed MDAC's TF is defined by charge conservation at nodes v_x and v^-. Solving the charge conservation equations, the differential output voltage can be shown to be

$$V_{od} = 2\left[1 + \frac{1}{4}\sum_{j=1}^{2}\left(+\frac{C_{p2j}}{C_{1j}} - \frac{C_{p6j}}{C_{1j}} - \frac{C_{p6j}}{C_{2j}} - \frac{C_{p2j}C_{p6j}}{C_{1j}C_{2j}} - \frac{C_{p3j}C_{p6j}}{C_{1j}C_{2j}}\right)\right]V_{id}. \tag{3.13}$$

The only parasitic capacitors that contribute to the GE are C_{p2j} and C_{p3j}. All others either affect the opamp's speed (C_{p1j}, C_{p7j}, and C_{p8j}) but not its accuracy, or have the same voltage between both phases (C_{p4j} and C_{p5j}). C_{p6j} is used for compensation. Equation 3.13 clearly shows that the MDAC is insensitive to capacitor mismatch because there are no ratio terms between main capacitors, but, on the other hand, is sensitive to parasitic capacitors C_{p2j}, C_{p3j}, and C_{p6j}. By analysing the signs of the terms multiplying V_{id}, it is evident that an appropriate value for C_{p6j} can compensate the term C_{p2j}/C_{1j}. More specifically, making $C_{1j} = C_{2j} = C$ and $C_{p2j} = C_{p3j} = C_p$, the term multiplying V_{id} is exactly 2 if $C_{p6j} = 0.5CCp/(C + Cp) = 0.5Cp/(1 + Cp/C)$. If $C_p \ll C$, then $C_{p6j} \approx 0.5C_p$. This result justifies the use of capacitor C_{3j}, since its top-plate parasitic capacitance is directly governed by C_{p6j} and parasitic compensation can be achieved by making $C_{3j} = 0.5C$ and by proper sizing of the MOS switches in the design phase and careful parasitic-aware layout.

To further explain the compensation issues involved, Fig. 3.7 will be used. This figure only shows half the MDAC circuit for simplicity, and adds names to the critical switches. Charge balancing occurs between two different circuit constructions, i.e.

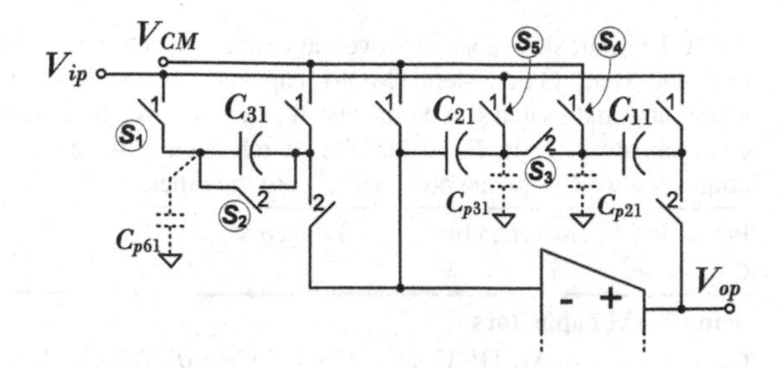

Fig. 3.7 Compensation circuit close-up and parasitic capacitor analysis

C_{21} and C_{11} circuitry relative to C_{31} circuitry. There could be concerns about certain assumptions made regarding the nature of the nodal parasitics. It is mentioned above that C_{p61} governs C_{31}'s top-plate parasitic capacitance (TPPC). However, C_{p61} is also governed by switches, S_1 and S_2, connected to its node. Also, it is assumed that $C_{p21} = C_{p31} = C_p$ but this requires balancing the bottom plate parasitic capacitance (BPPC) of C_{21} with the TPPC of C_{11} because of the series connection. Balancing TPPC and BPPC can be difficult to achieve with Metal-Insulator-Metal (MIM) capacitors, but is easier achieved with Metal-Oxide-Metal (MOM) capacitors (readily available in standard libraries of many CMOS technologies).[2] Balancing switches can be achieved by making S_1 and S_2 half of $S_{4,5}$ and S_3, respectively. Table 3.1 summarizes the effect and balancing of the crucial parasitic capacitors and demonstrates that $C_{p61} = C_{p21,31}/2$ can be achieved by using MOM capacitors, and careful design and layout (refer to Fig. 3.7 for component names). If MIM capacitors are employed, C_{p31} will not be compensated as desired. However, referring to Eq. 3.13, C_{p31} has a second order effect on the gain error and therefore, can be neglected.

The statistical analysis of the GE can be achieved by referring back to Eq. 3.13 and by defining $C_{ij} = C(1 + \varepsilon_{ij})$ and $C_{3j} = 0.5C(1 + \varepsilon_{ij})$, with $i, j = \{1, 2\}$, $C_{pij} = C_p(1 + \varepsilon_{pij})$ and $C_{p6j} = 0.5C_p(1 + \varepsilon_{p6j})/(1 + \alpha)$, with $i = \{2, 3\}$ and $j = \{1, 2\}$, where ε_{ij} and ε_{pij} are uncorrelated Gaussian random variables of the relative errors with zero mean and standard deviation σ, and $\alpha = C_p/C$. Monte Carlo simulations using MATLAB and Spectre are employed to evaluate the obtained expression. The standard deviation plus the absolute mean of the GE for different values of ε_{pij} and α, shown in Fig. 3.8a, proves that the GE is independent of capacitor mismatch, as expected. In Fig. 3.8b, the GE is plotted against the $\sigma(\varepsilon_{pij})$ for different values of α (for this simulation $\sigma(\varepsilon_{ij}) = 0.2\,\%$). It can be seen that the GE degrades for increasing α and it depends linearly on the $\sigma(\varepsilon_{pij})$. In addition, the GE continues compatible with resolutions of approximately 10 bits for $\alpha = 3\,\%$ and $\sigma(\varepsilon_{pij}) < 7\,\%$.

The combined effects of various α and parasitic capacitor mismatches is shown in Fig. 3.9. This graph shows iso-accuracy lines that represent the accuracy (in bits, for easier observability) of the GE. Through this graph it is possible to observe thc

[2] Notice that, by turning C_{21} around, the floating node has two top-plates connected to it, which is then easier to compensate. This comes at the expense of a reduction in the feedback factor because a larger parasitic capacitor (BPPC) is connected to the opamp's inputs.

Table 3.1 Parasitic capacitor governance and balancing using Metal-Insulator-Metal (MIM) capacitors and Metal-Oxide-Metal (MOM) capacitors. Capacitor sizes defined by $C_{11} = C_{21} = 2C_{31} = C$ and nominal switch size defined by W/L, where W and L are the width and length of the switch's channel, respectively. Note that, the switch contributes a parasitic capacitance to each node it is connected to: C_{gs} to one node and C_{gd} to the other, where $C_{gs} = C_{gd} = 0.5WLC_{ox}$

Parasitic Capacitor	Governed by	Balance
using MIM capacitors		
C_{p21}	S_3, S_4, $\text{TPPC}_{C_{11}}$	$C \times \text{TPPC}_C + 0.5WLC_{ox} + 0.5WLC_{ox} = C \times \text{TPPC}_C + WLC_{ox}$
C_{p31}	S_3, S_5, $\text{BPPC}_{C_{21}}$	$C \times \text{BPPC}_C + 0.5WLC_{ox} + 0.5WLC_{ox} = C \times \text{BPPC}_C + WLC_{ox}$
C_{p61}	S_1, S_2, $\text{TPPC}_{C_{31}}$	$\frac{C}{2} \times \text{TPPC}_C + 0.5\frac{W}{2} LC_{ox} + 0.5\frac{W}{2}LC_{ox} = \frac{C \times \text{TPPC}_C + WLC_{ox}}{2}$
using MOM capacitors		
C_{p21}	S_3, S_4, $\text{PPC}_{C_{11}}$	$C \times \text{PPC}_C + 0.5WLC_{ox} + 0.5WLC_{ox} = C \times \text{PPC}_C + WLC_{ox}$
C_{p31}	S_3, S_5, $\text{PPC}_{C_{21}}$	$C \times \text{PPC}_C + 0.5WLC_{ox} + 0.5WLC_{ox} = C \times \text{PPC}_C + WLC_{ox}$
C_{p61}	S_1, S_2, $\text{PPC}_{C_{31}}$	$\frac{C}{2} \times \text{PPC}_C + 0.5\frac{W}{2}LC_{ox} + 0.5\frac{W}{2}LC_{ox} = \frac{C \times \text{PPC}_C + WLC_{ox}}{2}$

PPC plate parasitic capacitance; *TPPC* top PPC; *BPPC* bottom PPC

allowed tolerances, in terms of parasitic capacitor and its mismatch, for a given GE accuracy. For these simulations capacitor mismatch, $\sigma(\varepsilon_{ij}) = 0.2\,\%$. Figure 3.9 shows a large area of values that achieve over 10-bit GE accuracy. Besides the GE accuracy, the CMRS-MDAC always benefits from a two-fold gain in the feedback factor, which will be shown shortly.

3.2.2.1 Opamp's Finite DC Gain

Considering the effect of the opamp's finite A_0, a more complete expression for the GE, and consequently C_{p6j}, is obtained. If C_{p6j} is adequately sized (and carefully laid out) it can also compensate for the opamp's finite A_0 and produce a highly accurate gain of two. The equations are too complex to show here, but the simulation results are shown in Fig. 3.10 and summarized in Table 3.2 for various values of $\overline{A_0}$ (and respective 3σ variations). In average, the proposed MDAC achieves an extra 2 bits in accuracy. These results show that for $\overline{A_0} = 60\,\text{dB}$ with $3\sigma[A_0] = 6\,\text{dB}$, more than 10-bit accuracy is achieved with the proposed MDAC.

3.2.2.2 Charge Injection and Clock Feed-Through

Regarding the effects of charge injection and clock feed-through on the compensation circuit, simulations show that these result in an offset term, as shown in

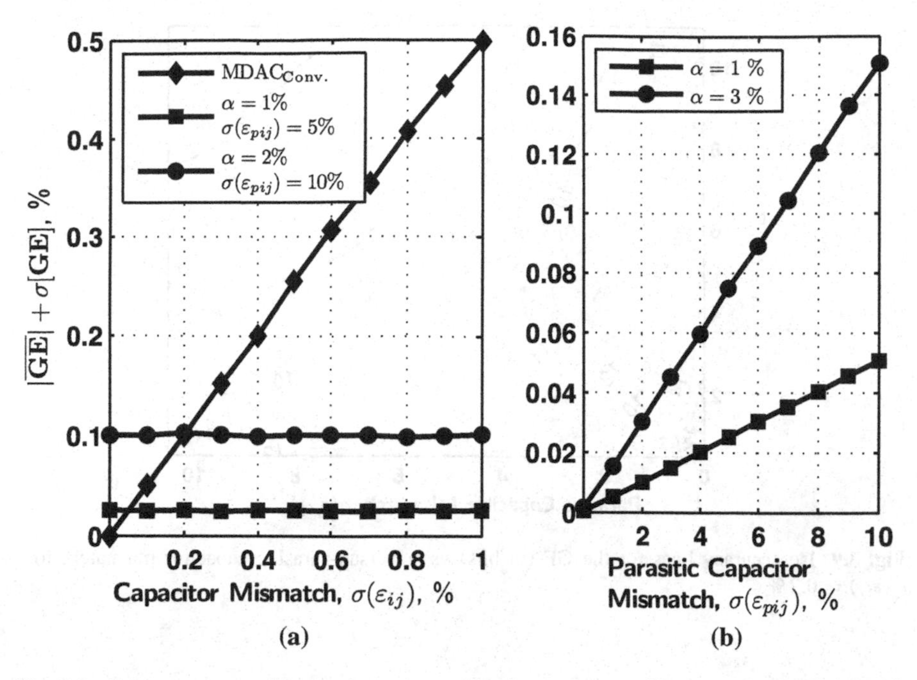

Fig. 3.8 Gain error versus **(a)** capacitor mismatch and comparison with conventional MDAC and **(b)** parasitic capacitor mismatch (for $\sigma(\varepsilon_{ij}) = 0.2\,\%$). Each data point is the result of 1000 Monte Carlo cases

$$V_{od} = 2\left[1 + \frac{1}{4}\sum_{j=1}^{2}\left(+\frac{C_{p2j}}{C_{1j}} - \frac{C_{p6j}}{C_{1j}} - \frac{C_{p6j}}{C_{2j}} - \frac{C_{p2j}C_{p6j}}{C_{1j}C_{2j}} - \frac{C_{p3j}C_{p6j}}{C_{1j}C_{2j}}\right)\right]V_{id} + q_{inj1}\left(\frac{C_{11} + C_{21} + C_{p21} + C_{p31}}{C_{11}C_{21}}\right) - q_{inj2}\left(\frac{C_{12} + C_{22} + C_{p22} + C_{p32}}{C_{12}C_{22}}\right), \tag{3.14}$$

where q_{inj1} and q_{inj2} represent the charge injection and clock feed-through of the positive and negative paths of the MDAC.[3] As can be seen neither nonideal effects affect the GE, given that the offset term is signal independent. The effect of charge injection is minimized with signal independent sampling which is assumed for these simulations, hence the offset term being signal independent. Nevertheless, V_{CM} (common-mode voltage or analog ground) dependent sampling occurs, which adds an offset to the output voltage. However, by using a fully differential design, the V_{CM} dependent charge injection will cancel at the differential output. Clock

[3] Charge injection is given by $q_{inj} = -0.5WLC_{ox}(V_G - V_T) + 0.5WLC_{ox}V_S$, where the V_S term is signal dependent. It is assumed that half the charge is injected into each node. Clock feed-through injected voltage is given by $V_G C_{ov} W/(C_{ov}W + C)$.

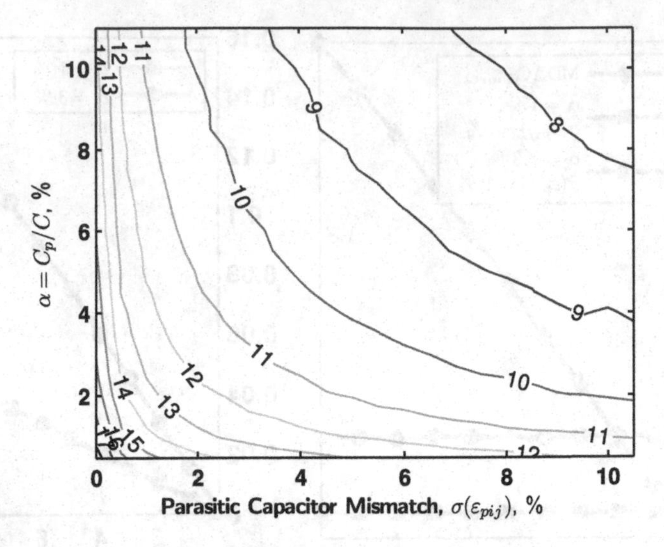

Fig. 3.9 Iso-accuracy lines for the GE (in bits) vs α versus parasitic capacitor mismatch, for $\sigma(\varepsilon_{ij}) = 0.2\,\%$

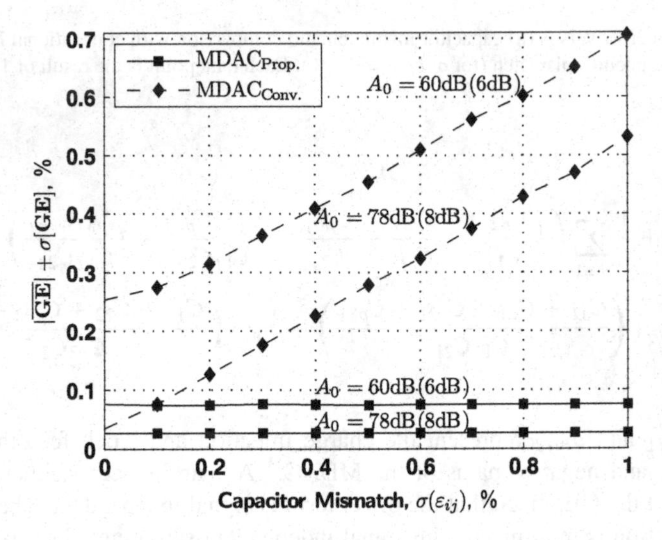

Fig. 3.10 Comparison between the conventional and the current-mode reference shifting MDAC for different opamp A_0 (and respective 3σ variations), for $\sigma(\varepsilon_{pij}) = 5\,\%$ and $\alpha = 1\,\%$. Each data point is the result of 1000 Monte Carlo cases

feed-through is attenuated by using dummy switches and, again, a fully differential design. Nevertheless, careful layout practice is important to minimize parasitic couplings and guarantee differential design. Figure 3.11a shows the offset for various

Table 3.2 Gain error (%) versus opamp's A_0 and 3σ variations (dB)($\sigma(\varepsilon_{ij}) = 0.2\,\%$, $\sigma(\varepsilon_{pij}) = 5\,\%$, and $\alpha = 1\,\%$)

$\overline{A_0}$ ($3\sigma[A_0]$)	40 (5)	50 (6)	60 (6)	70 (8)	80 (8)	90 (10)
GE_{CMRS}	0.59	0.23	0.076	0.04	0.026	0.025
$GE_{Conventional}$	2.4	0.83	0.32	0.17	0.12	0.11

values of charge injection and clock feed-through mismatch. This offset measures the mismatch between the charge injected in the two differential signal paths. For the simulation, capacitor and parasitic capacitor mismatch are considered.

3.2.2.3 V_{CM} Mismatch

By comparing Fig. 3.1 with Fig. 3.5a it can be noticed that V_{CM}, for the CMRS-MDAC is injected at two different nodes. The V_{CM} mismatch between these two nodes does not affect the accuracy of the gain of two. However, it does add an offset term to the MDAC's output voltage which can be minimized with fully differential design and careful V_{CM} path layout. Figure 3.11b shows the offset for various values of mismatch between the sampled V_{CM} of the two signal paths that should, under ideal conditions, cancel at the differential output. For the simulation, capacitor and parasitic capacitor mismatch are considered. As can be seen, for mismatches up to 100 %, the offset is negligible.

3.2.3 Reference Shifting Error Analysis

For the RE analysis, the circuit of Fig. 3.12 will be used. With $X = 1$, as an example, the current source I_N is connected to the inverting input of the opamp, while I_P is connected to the noninverting input. To simplify the circuit analysis, it will be assumed that the total current $I_k, k = \{P, N\}$ flows through the feedback path formed by the series associated capacitors C_{2j} and C_{1j}, $j = \{1, 2\}$, and the two switches that close the loop. This assumption is based on the fact that the remaining parasitic paths have much higher impedance and thus only a very small fraction of the current will flow through these paths. A fully differential analysis will be carried out for the RE, by defining $I_P = I_N = I_{REF}/2$, $R_1 = R_2 = 2R_{on}$, $C_j = C_{1j}C_{2j}/(C_{1j}+C_{2j})$, $j = \{1, 2\}$.[4] Defining the opamp's input nodes' voltages, V_p and V_n, as

[4] If I_P and I_N are not exactly matched, a current error (I_e) is introduced and results in an additive term appearing at the end of Eq. 3.18. It only affects the capacitor mismatch error (Eq. 3.20), depending
(Footnote 4 continued)
mainly on an I_e/I_{REF} term. Simulation results show that, for values of I_e/I_{REF} up to 20 %, this error may be neglected.

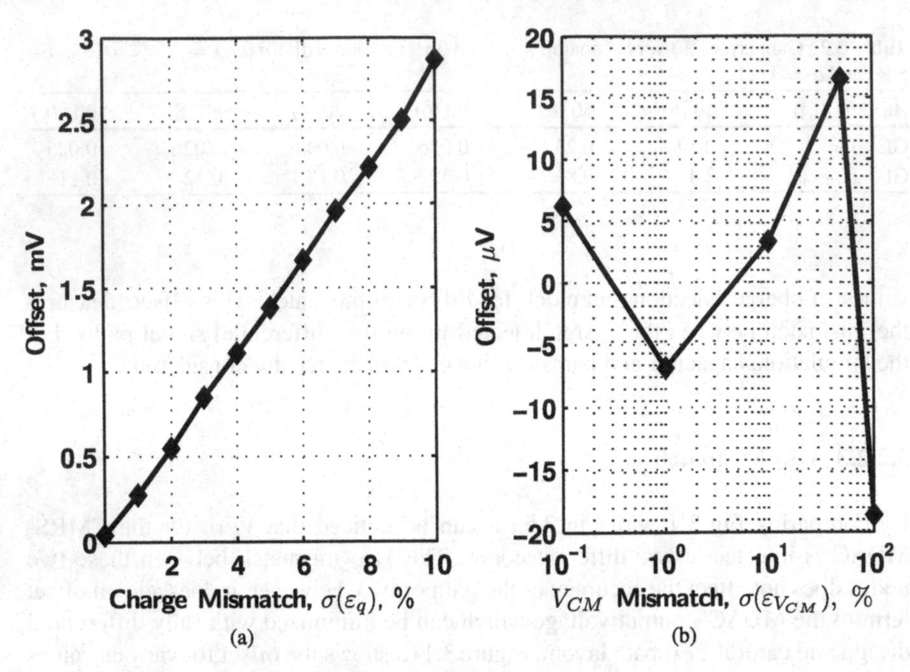

Fig. 3.11 Differential output offset due to: **a** Charge injection and clock feed-through mismatch and **b** V_{CM} mismatch. Simulation conditions for the capacitors: $\sigma_C(\varepsilon_{ij}) = 0.2\,\%$, $\sigma_p(\varepsilon_{pij}) = 5\,\%$, and $\alpha = 1\,\%$. For the switches: $W/L = 10/0.12$, gate voltage $V_G = 1.2\,$V, threshold voltage $V_T = 0.4\,$V, normalized gate oxide capacitance $C_{ox} = 12.6\,$fF/μm^2, and normalized overlap capacitance, $C_{ov} = 0.3\,$fF/μm. Each data point is the result of 1000 Monte Carlo cases

Fig. 3.12 MDAC configuration with current-mode reference shifting active for $X = 1$. Circuit used for RE analysis

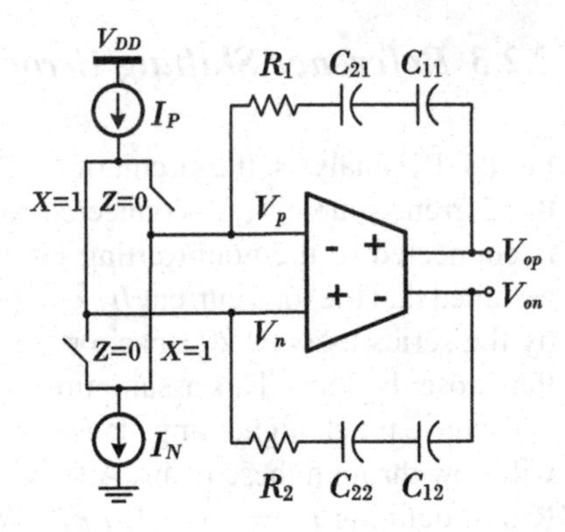

$$\begin{cases} V_p = V_{op} - (R_1 + 1/sC_1)I_{REF}/2 \\ V_n = V_{on} + (R_2 + 1/sC_2)I_{REF}/2, \end{cases} \tag{3.15}$$

and substituting them into $V_{od} = -A_v(V_p - V_n)$, solving for V_{od}, and applying the single pole TF of A_v, we obtain

$$V_{od}(s) = \frac{1}{2(1 + s/A_0 p_1)} \left(R_1 + R_2 + \frac{1}{sC_1} + \frac{1}{sC_2} \right) I_{REF}. \tag{3.16}$$

Considering a step function $I_{REF}(s) = I_{REF}/s$, $V_{od}(t)$ is obtained applying the inverse Laplace transform,

$$V_{od}(t) = \frac{I_{REF}}{2} \sum_{i=1}^{2} \frac{1}{C_i} \left[t + \frac{(e^{-\mathrm{GBW}t} - 1)(1 - \mathrm{GBW} R_i C_i)}{\mathrm{GBW}} \right], \tag{3.17}$$

where $\mathrm{GBW} = A_0 p_1$. Assuming GBW $\ll 1/(R_i C_i)$ and integration (or reference shifting) time $T_i = 1/(2F_S)$, we can simplify Eq. 3.17 to

$$V_{od}(T_i) \cong \frac{I_{REF}}{2} \left(T_i + \frac{e^{-\mathrm{GBW}T_i} - 1}{\mathrm{GBW}} \right) \left(\frac{C_{11} + C_{21}}{C_{11}C_{21}} + \frac{C_{12} + C_{22}}{C_{12}C_{22}} \right). \tag{3.18}$$

From Eq. 3.18 three sources of error emerge: opamp's dynamic limitation given by GBW, capacitor mismatch (ratio terms between capacitors) and integration time variations represented by T_i. These three sources of error are analysed next.

3.2.3.1 Opamp's Dynamic Limitation

Assuming $C_{11} = C_{21} = C_{12} = C_{22} = C$ and a fixed T_i, the error due to GBW is given by

$$\varepsilon_{\mathrm{GBW}} = \frac{V_{od} - V_{od_{ideal}}}{V_{od_{ideal}}} \cong \frac{-1}{\mathrm{GBW}T_i}, \tag{3.19}$$

where $V_{od_{ideal}} = 2 I_{REF} T_i / C$. As an example, a 12-bit application needs $N \ln(2) \approx 9$ time constants for linear settling. Therefore, the closed-loop $\mathrm{GBW} = 9 \times 2F_S = 9/T_i$ in units of rad/s. Substituting in Eq. 3.19 leads to an absolute error of 11 %. Although large, a small increase in I_{REF} compensates for this GBW limitation.

This compensation is demonstrated in Fig. 3.13. This graph combines various values of GBW and I_{REF}, including capacitor mismatch, with the final result being Eq. 3.18 (it is considered $V_{od} = V_{REF}$). The conditions for the simulations of this hypothetical case are: $C = 1\,\mathrm{pF}$, nominal GBW = 3.2 GHz, $T_i = 420\,\mathrm{ps}$, which yields a nominal $I_{REF} = 600\,\mu\mathrm{A}$. For capacitor mismatch it is considered $\sigma(\varepsilon_{ij}) = 0.2\,\%$. The objective is to achieve $V_{REF} = 0.5\,\mathrm{V}$. Figure 3.13 has two reference lines (dotted lines). The horizontal reference line for $I_{REF} = 600\,\mu\mathrm{A}$ shows that it is very difficult to obtain $V_{REF} = 0.5\,\mathrm{V}$ for practical values of GBW. On the other hand, the vertical reference line, for GBW = 3.2 GHz, shows that by increasing I_{REF} from 600 to 684 µA, $V_{REF} = 0.5\,\mathrm{V}$ is achieved. This corresponds to an increase of 14 % in the reference current. Therefore, a small increase in the reference current compensates for limited GBW.

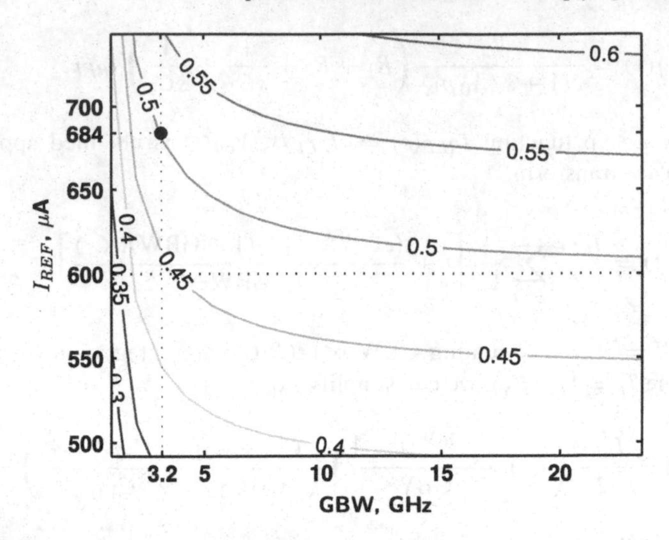

Fig. 3.13 Iso-V_{REF} lines versus GBW versus I_{REF}. Error due to finite GBW and compensation by increasing I_{REF}

3.2.3.2 Capacitor Mismatch

Assuming GBW $\rightarrow \infty$ and a fixed T_i, the resulting error is given by

$$\varepsilon_C = \frac{V_{od} - V_{od_{ideal}}}{V_{od_{ideal}}} = \frac{C}{4}\left(\frac{C_{11}+C_{21}}{C_{11}C_{21}} + \frac{C_{12}+C_{22}}{C_{12}C_{22}}\right) - 1. \tag{3.20}$$

If $C_{ij} = C(1+\varepsilon_{ij})$, where ε_{ij} are assumed uncorrelated Gaussian random variables with zero mean and standard deviation σ, and assuming σ sufficiently small, such that $1/(1+\varepsilon_{ij}) \approx 1 - \varepsilon_{ij}$, we obtain $\sigma_{\varepsilon_C} = \sigma_C/2 = 0.1\,\%$ (assuming $\sigma_C = 0.2\,\%$). Therefore, this error is easily corrected by the digital correction logic.

3.2.3.3 Integration Time Variation

Assuming GBW $\rightarrow \infty$, all capacitors are equal, and an integration time given by $T_i(1+\varepsilon_{T_i})$, where ε_{T_i} represents the deviation from the ideal integration time, the resulting integration time error is

$$\varepsilon_T = \frac{V_{od} - V_{od_{ideal}}}{V_{od_{ideal}}} = \frac{T_i(1+\varepsilon_{T_i}) - T_i}{T_i} = \varepsilon_{T_i}. \tag{3.21}$$

This means that this error is dependent on the jitter noise of the system the MDAC is embedded in. Therefore, comparing all error sources, this represents the smallest one, which are dominated by the finite GBW of the opamp.

A final TF expression for the CMRS-MDAC can be obtained by summing Eq. 3.13 with B times Eq. 3.18 (B represents the MDAC operation mode), which is given by

$$\begin{aligned} V_{od} =& 2\left[1+\frac{1}{4}\sum_{j=1}^{2}\left(+\frac{C_{p2j}}{C_{1j}}-\frac{C_{p6j}}{C_{1j}}-\frac{C_{p6j}}{C_{2j}}-\frac{C_{p2j}C_{p6j}}{C_{1j}C_{2j}}-\frac{C_{p3j}C_{p6j}}{C_{1j}C_{2j}}\right)\right]V_{id} \\ &+B\times\frac{1}{2}\left(T_i+\frac{e^{-\mathrm{GBW}T_i}-1}{\mathrm{GBW}}\right)\left(\frac{C_{11}+C_{21}}{C_{11}C_{21}}+\frac{C_{12}+C_{22}}{C_{12}C_{22}}\right)I_{REF}. \quad (3.22)\end{aligned}$$

3.2.4 Feedback Factor

The β of the CMRS-MDAC can be derived with the help of Fig. 3.5c considering C_{p71} (i.e., the opamp's input parasitic capacitance, see Fig. 3.5a) as the dominant parasitic capacitor, which results

$$\beta_{\mathrm{Y}}=\frac{V^{-}(s)}{V_o(s)}=\frac{1}{1+C_{p71}\frac{C_{11}+C_{21}}{C_{11}C_{21}}}. \quad (3.23)$$

If $C_{p71} \ll C_{11}C_{21}/(C_{11}+C_{11})$, the feedback factor approximates unity. Therefore, the resulting β is two times greater than that of the conventional MDAC, which is clearly a relevant advantage since the speed/power ratio doubles. The enhanced β also reduces the effective load, $C_{Leff} = C_L + (1-\beta)C_{feedback}$ ($C_{feedback}$ is the equivalent series feedback capacitance), therefore the total gain is approximately 2.5 times over the conventional MDAC.[5] This can be seen in the simulation results of Fig. 3.14, where the closed-loop time constant and 12-bit settling time of the conventional MDAC are 2.5 times that of the proposed MDAC. For these simulations the following conditions are considered: the model of the opamp has $A_0 = 106$ dB, GBW = 3.2 GHz, and a load capacitance, $C_L = 2$ pF.

Equation 3.23 refers to the feedback factor for Y operation mode. In X and Z modes, considering a finite output impedance of the current source (an R_S in parallel with a C_S), the feedback factor is given by

$$\beta_{\mathrm{X,Z}}=\frac{1}{1+\frac{C_{11}+C_{21}}{sR_SC_{11}C_{21}}+(C_S+C_{p71})\frac{C_{11}+C_{21}}{C_{11}C_{21}}}. \quad (3.24)$$

For high output resistance ($R_S \to \infty$), the second denominator term becomes negligible and for small C_S or for an adequately saturated current source (C_S has very little effect), Eq. 3.24 becomes similar to Eq. 3.23.

[5] This is true when comparing both the conventional and CMRS-MDACs with an equal load capacitor, C_L. However, in a pipeline (or similar) ADC, employing the CMRS-MDAC in all its stages, the β enhancement reduces to 2, because of the extra sampling half-capacitor, C_{3j}, which loads each stage.

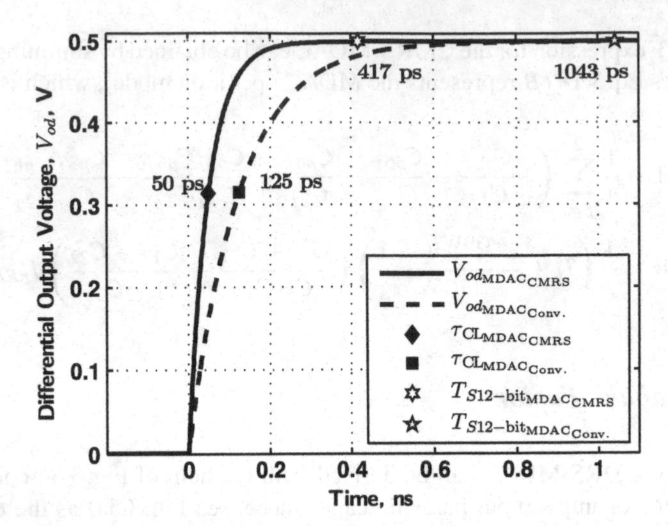

Fig. 3.14 Step response comparison between conventional MDAC and CMRS-MDAC, illustrating the effect of the β enhancement

3.2.5 Thermal Noise Analysis

Figure 3.15 shows the equivalent single-ended circuits for noise analysis of the CMRS-MDAC. The sampled noise, the noise from switch $v_{SW,2}$, and the opamp's noise are identical to those of the conventional MDAC (see Eq. 3.12). C_{3j} do not accumulate noise because they are shorted in ϕ_2. Therefore it is only necessary to determine the noise contributions of $v_{SW2,2}$ and i_{n1}.

Starting with the noise contribution of $v_{SW2,2}$, assuming that $C_{eq} = C_{11}C_{21}/(C_{11}+C_{21})$ and that the switches' ON-resistances are negligible, we can write the noise TF as

$$H_{SW2,2}(s) = \frac{v_o}{v_{SW2,2}} = \frac{1/\beta}{[1+s/(\beta A_0 p_1)][1+sR_S(C_{eq}+C_{p71})]}, \tag{3.25}$$

where $\beta = 1/(1+C_{p71}/C_{eq})$. Consequently the MS output noise due to the switches' ON-resistance is given by

$$\overline{v^2_{osw2,2}} = 4kTR_{on}\left(\frac{1}{4\beta^2}\frac{\beta A_0 p_1}{1+\beta A_0 p_1 R_S(C_{eq}+C_{p71})}\right) = \frac{kTR_{on}}{\beta^2 R_S(C_{eq}+C_{p71})}. \tag{3.26}$$

where $R_S(C_{eq}+C_{p71}) \gg 1/(\beta A_0 p_1)$ is assumed.

The TF of the last noise source, i_{n1}, is given by

$$H_{in1}(s) = \frac{v_o}{i_{n1}} = \frac{A_v}{(A_v+1)sC_{eq}} = \frac{1}{(1+s/(A_0 p_1))sC_{eq}}. \tag{3.27}$$

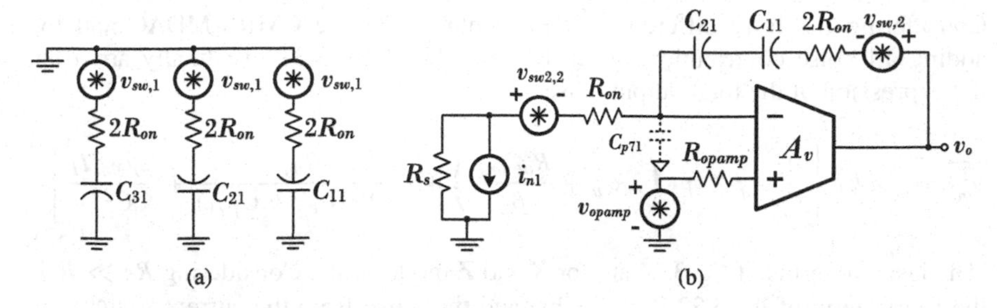

Fig. 3.15 Equivalent circuits for thermal noise analysis of the CMRS-MDAC in: **a** ϕ_1 and **b** ϕ_2

In the derivation of Eq. 3.27 it is assumed that all noise current flows through the feedback path and that $A_0 \gg 1$. We note that this TF has a pole at zero, which resembles an integrating characteristic, and thus, the noise PSD integral is unbounded at the low frequency limit. Therefore, the method used until now to compute the output noise can not be directly applied[6] [151]. An alternative method uses a rectangular window, $w(t)$, from 0 to T_i (integration time), to shape the impulse response $h_{in1}(t)$, i.e., $h_{in1w}(t) = h_{in1}(t)w(t)$, and then applies the traditional method with the TF H_{in1w} [151]. The impulse response obtained from Eq. 3.27 is

$$h_{in1}(t) = \frac{1 - e^{-A_0 p_1 t}}{C_{eq}}. \tag{3.28}$$

Applying a rectangular window function, we have

$$h_{in1w}(t) = h_{in1}(t)w(t) = h_{in1}(t)[u(t) - u(t - T_i)], \tag{3.29}$$

where $u(t)$ is the unit step function. By determining the Fourier transform of Eq. 3.29 and knowing that the PSD of the noise current is $S_{in1}(f) = 4\gamma kTg_m$ (where γ and g_m are the transistor's excess noise factor and transconductance, respectively), the contribution to the output noise can be calculated as

$$\overline{v_{oin1}^2} = \int_0^{\infty} |H_{in1w}(j2\pi f)|^2 S_{in1}(f)\mathrm{d}f = 4\gamma kTg_m \left(\frac{-3 + 2a + 4e^{-a} - e^{-2a}}{4C_{eq}^2 A_0 p_1}\right), \tag{3.30}$$

where $a = A_0 p_1 T_i = \mathrm{GBW}T_i$. For $\mathrm{GBW} \gg 1/T_i$, Eq. 3.30 can be simplified to

$$\overline{v_{oin1}^2} = \frac{2\gamma kTg_m T_i}{C_{eq}^2}. \tag{3.31}$$

---

[6] The method used until now is based on the formula $\int_0^{\infty} |H(j2\pi f)|^2 S(f)\,df$.

Considering the fully differential implementation of the CMRS-MDAC and by adding all noise contributions given by Eq. 3.12, 3.26, 3.31, we finally arrive at the expression of the total output noise,

$$\overline{v_{od}^2} = 2 \times kT \left[\frac{2}{C} + \beta A_0 p_1 \left(2R_{on} + \frac{R_{opamp}}{\beta^2} \right) + \frac{R_{on}}{\beta^2 R_S (C_{eq} + C_{p71})} + \frac{2\gamma g_m T_i}{C_{eq}^2} \right]. \tag{3.32}$$

The last two terms of Eq. 3.32 are for X and Z modes only. Considering $R_S \gg R_{on}$ the fourth term of Eq. 3.32 is negligible and the noise from the current source can be minimized by increasing the unit capacitance of each main capacitor. Comparing Eq. 3.32 with Eq. 3.12 at first sight seems that the noise of the proposed MDAC is larger. However, looking at Y mode operation shows that the conventional MDAC's noise is larger given its lower feedback factor, which increases the noise of the opamp. For X and Z modes, the noise of the proposed MDAC is larger mainly due to the noise of the current sources.

This concludes the analysis of the current-mode reference shifting MDAC circuit. The following section analyses the sampling phase reference shifting MDAC.

3.3 Sampling Phase Reference Shifting MDAC

This section proposes a second MDAC circuit with unity feedback factor, insensitive to capacitor mismatch, where the main difference compared to the previous MDAC is the reference shifting scheme. This scheme consists of voltage-mode reference shifting, which occurs immediately during the sampling phase, unlike the conventional MDAC and the CMRS-MDAC, where it occurs during the amplification phase. As an abbreviation this MDAC will be called SPRS-MDAC.

3.3.1 Principle of Operation

The enhanced feedback factor 1.5-bit fully differential MDAC is shown in Fig. 3.16. The circuit operates in two clock phases but it requires valid outputs from the local quantizer (X, Y, and Z) still during ϕ_1. In ϕ_1, the differential input voltage is sampled on capacitors C_{1j}, C_{2j}, and C_{3j} for the positive ($j = 1$) and negative ($j = 2$) signal paths. Also, a reference shifting voltage ($-V_{REF}, 0, +V_{REF}$) dependent on X, Y, and Z is sampled on capacitors C_{1j}. During ϕ_2, capacitors C_{1j} and C_{2j} are associated in series and the MDAC's output characteristic (see Eq. 3.1) is obtained by voltage sum, instead of charge redistribution, as in the conventional implementation. As a result, this circuit is inherently insensitive to capacitor mismatches and nonlinearities.

Like the previous MDAC, this approach for obtaining a capacitor mismatch-insensitive gain of two causes the circuit to become sensitive to parasitic capacitors C_{p2j} and C_{p3j} (nodes between C_{2j} and C_{1j}). To attenuate this sensitivity, capaci-

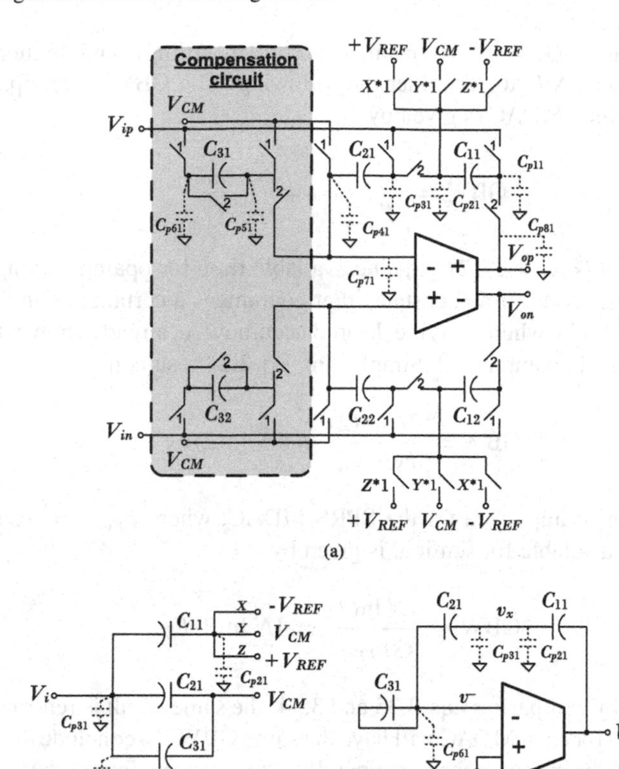

Fig. 3.16 **a** Enhanced feedback factor, fully differential 1.5-bit mismatch-insensitive MDAC with sampling phase reference shifting. **b** Single-ended equivalent circuit during sampling (ϕ_1). **c** Single-ended equivalent circuit during ϕ_2 for GE analysis

tors C_{3j} are employed. Notice, however, that these capacitors are shorted during ϕ_2 because only the charge stored in their parasitic capacitors (C_{p6j}) is used to compensate the charge stored on C_{p2j} and C_{p3j}. The compensation circuit is shown in the shaded area of Fig. 3.16a.

This MDAC is capable of achieving a feedback factor of one, is insensitive to capacitor mismatch, and still uses voltage-mode reference shifting. All these benefits are accompanied by some penalties, namely the opamp's dynamics. As mentioned above, this MDAC needs the local quantizer's decision during ϕ_1, which means that the local quantizer has half the sampling period to make a decision. After that decision is made, in case it is X or Z, $\pm V_{REF}$ has to be switched in at C_{11}'s top-plate. This switching causes a disturbance of the sampling, and as a result, the dynamics of the MDAC changes and the opamp needs to re-settle. To avoid prolonged settling (after $\pm V_{REF}$ has been switched in), the speed of the opamp has to be higher. To demonstrate that the GBW of this opamp (for correct settling) is not higher than that of the conventional MDAC, consider the following. The GBW of an opamp sized for the conventional MDAC is given by

$$\mathrm{GBW} = \frac{1}{\tau\beta}, \qquad \tau = \frac{T_{settling}}{n} \tag{3.33}$$

where $T_{settling} = T_S/2 = 1/(2F_S)$ is the available time for opamp settling and n is the number of necessary time constants that guarantees a certain settling accuracy, given by $n = N\ln(2)$, where N is the desired accuracy. As already shown, the $\beta_{\mathrm{Conv.}}$ of the conventional opamp is 1/2. Simplifying Eq. 3.33 results in

$$\mathrm{GBW} = \frac{2N\ln(2)}{T_S\beta_{\mathrm{Conv.}}} = 4N\ln(2)F_S \tag{3.34}$$

The GBW of an opamp sized for the SPRS-MDAC, where $\beta_{\mathrm{Prop.}} = 1$ and only a quarter of T_S is available for settling, is given by

$$\mathrm{GBW} = \frac{4N\ln(2)}{T_S\beta_{\mathrm{Prop.}}} = 4N\ln(2)F_S \tag{3.35}$$

As can be seen by comparing Eqs. 3.35 and 3.34, the same result is rendered. Therefore, the opamps for each MDAC will have the same GBW. To conclude this analysis, although the feedback factor enhancement does not translate into higher power efficiency, this MDAC still benefits from its insensitivity to capacitor mismatch, as already demonstrated.

3.3.2 Gain Error Analysis

Detailed gain error analysis for this MDAC in Y mode has already been carried out in the gain error analysis subsection of the previous MDAC (Sect. 3.2.2), and therefore, will not be repeated here.

3.3.3 Reference Shifting Error Analysis

This MDAC's TF is defined by charge conservation at nodes v_x and v^-, similar to the previous MDAC. Solving the charge conservation equations, the differential output voltage with reference shifting is given by

$$V_{od} = 2\left[1 + \frac{1}{4}\sum_{j=1}^{2}\left(+\frac{C_{p2j}}{C_{1j}} - \frac{C_{p6j}}{C_{1j}} - \frac{C_{p6j}}{C_{2j}} - \frac{C_{p2j}C_{p6j}}{C_{1j}C_{2j}} - \frac{C_{p3j}C_{p6j}}{C_{1j}C_{2j}}\right)\right]V_{id}$$

$$+\left(1 + \frac{1}{2}\sum_{j=1}^{2}\frac{C_{p2j}}{C_{1j}}\right)B \cdot V_{REF}, \quad (3.36)$$

where V_{REF} represents the differential reference voltage. As can be seen from Eq. 3.36 the term multiplying V_{id} is the same as that shown in Eq. 3.13, while the term multiplying V_{REF} represents the reference shifting term, which should ideally be one. Therefore, the reference shifting error is given by $1/2 \sum_{j=1}^{2} C_{p2j}/C_{1j}$, which is of the order C_p/C, thus is not of great concern since it can be corrected by the digital correction logic.

3.3.3.1 Charge Injection and Clock Feed-Through

The effects of charge injection and clock feed-through are more pronounced in this MDAC due to the V_{REF} sampling. Sampling capacitors $C_{1j}, j = \{1, 2\}$, depending on the MDAC's operation mode, may sample $\pm V_{REF}$ in one signal path, while the other signal path samples $\mp V_{REF}$. This means that the injected charge will not cancel at the differential output, even under ideal conditions. Rewriting the charge conservation equations considering V_{REF} charge injection and clock feed-through, the MDAC's TF becomes

$$V_{od} = 2\left[1 + \frac{1}{4}\sum_{j=1}^{2}\left(+\frac{C_{p2j}}{C_{1j}} - \frac{C_{p6j}}{C_{1j}} - \frac{C_{p6j}}{C_{2j}} - \frac{C_{p2j}C_{p6j}}{C_{1j}C_{2j}} - \frac{C_{p3j}C_{p6j}}{C_{1j}C_{2j}}\right)\right]V_{id}$$

$$+\left(1 + \frac{1}{2}\sum_{j=1}^{2}\frac{C_{p2j}}{C_{1j}}\right)B \cdot V_{REF}$$

$$+ q_{inj1}\left(\frac{C_{11} + C_{21} + C_{p21} + C_{p31}}{C_{11}C_{21}}\right) - q_{inj2}\left(\frac{C_{12} + C_{22} + C_{p22} + C_{p32}}{C_{12}C_{22}}\right)$$

$$+ \frac{q_{vref+}}{C_{11}} - \frac{q_{vref-}}{C_{12}}, \quad (3.37)$$

where the first line represents the gain of two, the second line is the reference shifting term, the third line represents the charge injection (gate voltage term) and clock feed-through, and finally, the last line shows the charge injected due to the different sampled reference voltages (source voltage term). As should be noticed, the reference shifting error also contributes to the total offset, which is the largest offset contributor. The total offset is shown in Fig. 3.17 for various values of charge injection and clock feed-through mismatch. Capacitor and parasitic capacitor mismatch are

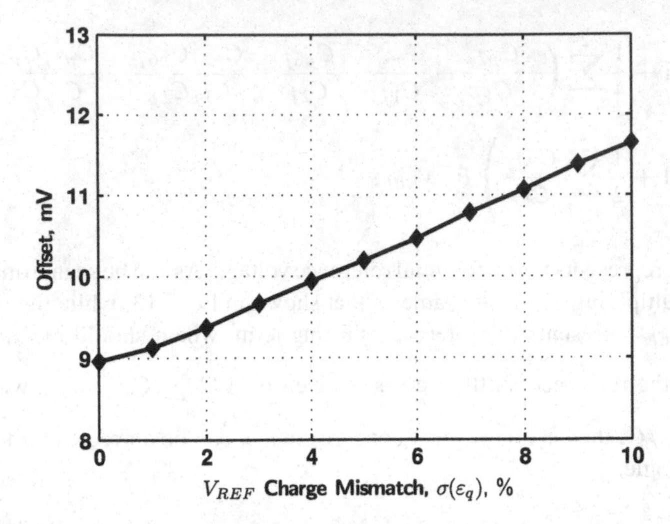

Fig. 3.17 Differential output offset due to V_{REF} charge injection and clock feed-through mismatch. Simulation conditions for the capacitors: $\sigma_C(\varepsilon_{ij}) = 0.2\,\%$, $\sigma_p(\varepsilon_{pij}) = 5\,\%$, and $\alpha = 1\,\%$. For the switches: $W/L = 10/0.12$, gate voltage $V_G = 1.2\,\text{V}$, threshold voltage $V_T = 0.4\,\text{V}$, differential $V_{REF} = 0.5\,\text{V}$, normalized gate oxide capacitance $C_{ox} = 12.6\,\text{fF}/\mu\text{m}^2$, and normalized overlap capacitance, $C_{ov} = 0.3\,\text{fF}/\mu\text{m}$. Each data point is the result of 1000 Monte Carlo cases

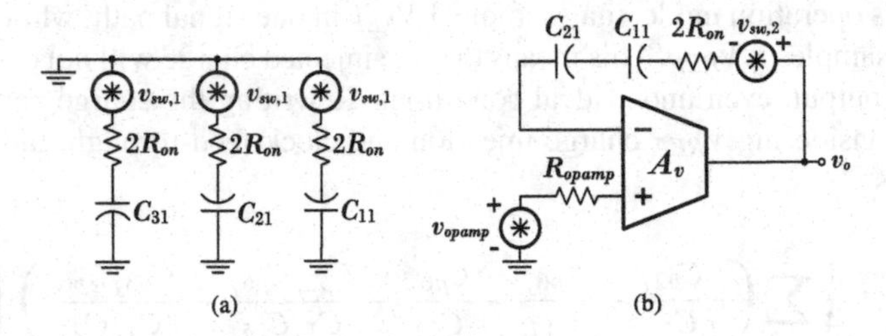

Fig. 3.18 Equivalent circuits for thermal noise analysis of the SPRS-MDAC in: **a** ϕ_1 and **b** ϕ_2

also considered. As can be seen in Fig. 3.17, even for matched charge injection, the offset error is around 9 mV (a normalized full-scale input of 1 Vpp is assumed). Nevertheless, this error can still be corrected by the DCL.

3.3.4 Feedback Factor

The feedback factor analysis is identical to the previous MDAC circuit and the β expression is described by Eq. 3.23 (see Sect. 3.2.4). As mentioned before, neglecting parasitics, it is nominally equal to 1.

3.3.5 Thermal Noise Analysis

For the thermal noise analysis of this MDAC, the equivalent circuits shown in Fig. 3.18 are used. Excluding C_{3j} $(j = \{1, 2\})$ noise accumulation from the analysis, given that it is shorted during ϕ_2, the noise analysis becomes identical to the one carried for the conventional MDAC (Sect. 3.1.4). Therefore only the final noise expression, for the fully differential implementation, is repeated here for convenience,

$$\overline{v_{od}^2} = 2 \times kT \left[\frac{2}{C} + \beta A_0 p_1 \left(2R_{on} + \frac{R_{opamp}}{\beta^2} \right) \right]. \tag{3.38}$$

3.4 Performance Summary and Comparison of 1.5-bit MDACs

A performance summary of the key characteristics of the current-mode (CMRS) and sampling phase reference shifting (SPRS) MDACs is shown in Table 3.3. The table also compares these two circuits with the conventional MDAC, and other capacitor mismatch-insensitive MDACs and multiply-by-two amplifiers (MBTAs) found in the literature from the past 10 years.

Regarding the data presented in Table 3.3, there is only one other structure that achieves a unity β [138]. It is important to notice that this MDAC achieves unity β at the cost of using two opamps (possibly doubling the static power consumption of the MDAC). In terms of gain error performance, it is difficult to evaluate and compare the data because each circuit is simulated for different conditions and some references do not present these conditions. Nevertheless, of the presented data, [189] achieves the best gain accuracy but uses a very small value of σ_p, and [95, 138] achieve a very good gain accuracy too, but [138] uses two opamps and [95] employs a complicated timing scheme, that could result in an output modulation of the input signal.

Regarding effective load ($C_{Leff} = C_L + (1 - \beta)C_{feedback}$), circuits with $\beta = 1$ achieve the best results. However, the results of this column must be observed with some caution when considering a pipeline (or similar) ADC architecture, where all stages use the same circuit structure. In this situation, C_L is substituted by the respective values of the 'Input Sampling Capacitance' column. Using as an example the conventional and proposed MDACs, the effective load of the conventional MDAC (if it is employed in all stages of a pipeline converter) is $2.5C$ which is the same for the proposed MDACs, given their higher input sampling capacitance.

Concerning the number of necessary phases to execute circuit operation, most rely on a 2-phase operation, while others need 3-phases or even 4-phases to complete the gain of two (with reference shifting). Regarding circuit complexity, the conventional MDAC is the simplest circuit to implement, while others employ a huge number of switches [87, 144, 188, 190] and capacitors [87, 144], some need two opamps [25, 87, 138], and [190] employs a four-input opamp.

Table 3.3 Key performance summary of the proposed MDACs and comparison with other capacitor mismatch-insensitive MDAC and MBTA circuits (from the past decade)

Ref.	β	kT/C Noise	Capacitor mismatch insensitive	V_{REF} buffers required	X, Y, Z required in	Gain error (%)	3σ Relative gain mismatch	Effective load[a]	Input sampling capacitance	Number of phases	Hardware complexity (SW,C,OA)[b]
MDAC$_{CMRS}$	**1**	$4kT/C$	✓	×	ϕ_2	**0.026**[c]	$3\frac{\alpha\sigma_p}{2}$	C_L	2.5C	**2**	**(26,6,1)**
MDAC$_{SPRS}$	**1**	$4kT/C$	✓	✓	ϕ_1	**0.026**[c]	$3\frac{\alpha\sigma_p}{2}$	C_L	2.5C	**2**	**(24,6,1)**
Conv.[98]	1/2	$4kT/C$	×	✓	ϕ_2	0.1[c]	$3\frac{\sigma_C}{2}$	$C_L + \frac{1}{2}C$	2C	2	(13,4,1)
[138]	1	$4kT/C$	✓	✓	ϕ_2	0.004	$3\frac{\alpha(\sigma_C+\sigma_p)}{2\sqrt{2}}$	C_L	2C	2	(16,4,2)
[144]	1/13	$4kT/C$	✓	✓	ϕ_2	N/A	N/A	$C_L + \frac{1}{2}C$	C	3	(42,12,1)
[25]	1/2	$7kT/C$	✓	✓	ϕ_2, ϕ_3	N/A	N/A	$C_L + \frac{2}{3}C$	2C	3	(–,8,2)
[190]	1/3	N/A	✓	✓	ϕ_1	0.014[f]	$3\alpha\frac{\sqrt{2}}{2}\sigma_p$	$C_L + \frac{2}{3}C$	C_{ip}	3	(36,6,1)
[29, 30]	1/2	N/A	✓	✓	ϕ_3, ϕ_4	0.006[e]	$3\sqrt{3\sigma_C^2}$	$C_L + \frac{1}{2}C$	C	4	(–,4,1)
[87]	1/2	N/A	✓	✓	ϕ_3	0.02	N/A	$C_L + \frac{1}{2}C$	2C	4	(54,20,2)
Multiply-by-two amplifiers (MBTAs)											
[56]	1/3	$12kT/C$	✓	—	—	0.025[c]	$3\frac{2\alpha\sigma_p}{5}$	$C_L + \frac{1}{3}C$	3C	2	(30,8,1)
[189]	1/3	$2kT/C$	✓	—	—	0.0021[d]	N/A	$C_L + \frac{2}{3}C$	C	2	(20,6,1)
[95]	1/2	$2kT/C$	✓	—	—	0.005[g]	$3\frac{1}{4}\frac{\sigma_C^2}{1+\sigma_C}$	$C_L + \frac{1}{2}C$	C	4	(28,6,1)
[188]	1/2	N/A	✓	—	—	0.0075[g]	$3\frac{\alpha\sigma_p}{2}$	$C_L + \frac{2}{3}C$	C	3	(36,6,1)

[a] $C_{Leff} = C_L + (1-\beta)C_{feedback}$, where C_L is a load capacitance [b] SW number of switches, C number of capacitors, and OA number of opamps [c] $\sigma_C(\varepsilon_{ij}) = 0.2\,\%$, $\sigma_p(\varepsilon_{pij}) = 5\,\%$, $\alpha = 1\,\%$, $\overline{A_0} = \infty$ [d] $\sigma_C(\varepsilon_{ij}) = 0.3\,\%$, $\sigma_p(\varepsilon_{pij}) = 0.3\,\%$, $\alpha = 1\,\%$, $\overline{A_0} = \infty$ [e] $\sigma_C(\varepsilon_{ij}) = 0.5\,\%$ [f] $\sigma_p(\varepsilon_{pij}) = 10\,\%$ [g] $\sigma_C(\varepsilon_{ij}) = 1\,\%$

Chapter 4
Application of Circuit Enhancement Techniques to ADC Building Blocks

Abstract In the first section of this chapter, a 1.5-bit flash quantizer is proposed. This fully differential flash quantizer has built-in thresholds, made possible by employing inverter structures as input devices. Self-biasing techniques are employed for enhanced PVT robustness. Various analyses (confirmed with simulations) are carried out to describe the circuit's functionality, such as, kickback noise, regeneration time, metastability, offset, sensitivity to common-mode variations, and finally, a working proof of a pipeline ADC that employs the proposed circuit in all stages is given. A design procedure is also described and the section is concluded with a performance summary and a comparative table. The second section of this chapter presents a two-stage amplifier with enhanced performance. Energy efficiency is improved by using inverter-input structures, which effectively double the transconductance of the circuit for the same current. Self-biasing is employed in both stages for improved PVT robustness and further power reduction. The analyses carried out include differential-mode and common-mode feedback, noise, offset, slew rate, input-output ranges, and some considerations are given what concerns the amplifier's class of operation. Finally, guidelines are given for a successful design and a genetic algorithm optimization procedure is briefly described.

4.1 Inverter-Based Self-Biased 1.5-bit Flash Quantizer

Conventional 1.5-bit quantizers employ 2 comparators followed by a binary XYZ encoder. Normally each comparator has an input switched-capacitor (SC) network, which defines the threshold voltage ($V_{TH} = \pm V_{REF}/4$), followed by a dynamic pre-amplifier stage (pre-amp), a dynamic positive feedback latch (PFBL), and an SR digital latch. In advanced CMOS technologies with low supply voltages, the use of the pre-amp attenuates metastability problems (caused by the reduced signal swing). Large devices are used in the pre-amp and PFBL to reach the required accuracy (offset), leading to a poor energy efficiency. In [122], for example, the pre-amp stage

M. Figueiredo et al., *Reference-Free CMOS Pipeline Analog-to-Digital Converters*, Analog Circuits and Signal Processing, DOI: 10.1007/978-1-4614-3467-2_4,

is embedded into the SC network by employing MOS parametric amplification, but an input SC network is still required. The input SC network can be avoided if built-in threshold levels are created by threshold inverter quantization [94] or by using quantum voltage comparators [186]. Although the techniques in [94] and [186] can be readily applied in low-medium resolution flash quantizers [139, 185], they are unsuitable for pipeline ADCs, since only a single-ended input can be accepted. Moreover, the circuits based on these techniques are highly sensitive to process, supply, and temperature (PVT) variations and mismatch.

In the proposed flash quantizer, the improvement of energy efficiency is achieved by proposing a fully differential 1.5-bit quantizer that relies on a simple circuit based on inverters with built-in threshold voltages. Self-biasing together with the proposed design procedure makes the circuit highly robust to PVT variations and mismatch. The generalization for sub-ADCs with higher resolutions (up to 3.5 or 4 bits) is straightforward, since any pair of symmetrical threshold voltages is easily built. Although the main application of this circuit targets pipeline ADCs, it is directly applicable in other types of A/D architectures, namely, multi-stage algorithmic and multi-step flash.

Simulation results, in standard 65 nm and 0.13 μm CMOS technologies, demonstrate that following the suggested design methodology, the proposed flash quantizer is capable of achieving low offset, low kickback noise, low metastability probability errors and fast regeneration time with very low power dissipation, leading to an enhancement in energy efficiency.

4.1.1 Principle of Operation

The 1.5-bit flash quantizer, illustrated in Fig. 4.1a, has three main inverters. INV_1 and INV_2 are connected to the differential input signal while INV_3, the self-biasing inverter, has at its input a common mode voltage (V_{CM}). The differential circuit composed of INV_1 and INV_3, defines one threshold voltage. Likewise, INV_2 and INV_3 form another differential circuit, which define the symmetric threshold voltage. Assuming that the input common-mode voltage is approximately V_{CM}, it is possible to extract the following relationships

$$\begin{cases} V_{GSN1} - V_{GSN3} + V_{DS4A} - V_{DS4B} = +V_{id}/2 \\ V_{SGP1} - V_{SGP3} = -V_{id}/2 \end{cases} \quad (4.1a)$$

$$\begin{cases} V_{GSN2} - V_{GSN3} + V_{DS4A} - V_{DS4B} = -V_{id}/2 \\ V_{SGP2} - V_{SGP3} = +V_{id}/2 \end{cases} \quad (4.1b)$$

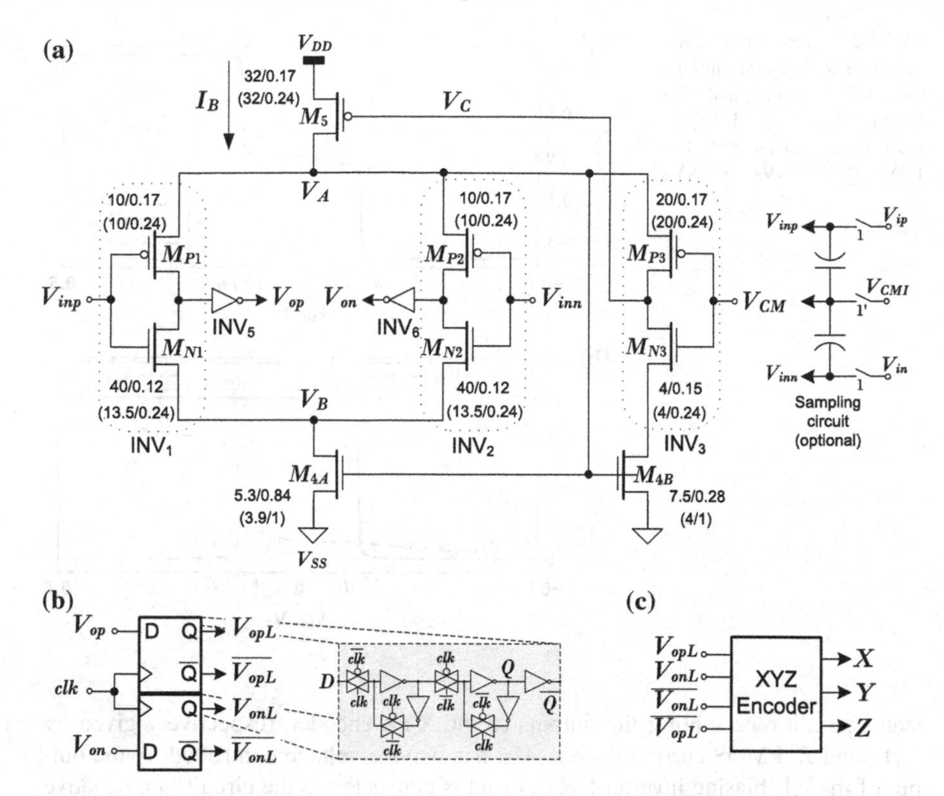

Fig. 4.1 Proposed 1.5-bit flash quantizer: **a** complete circuit with optional sampling circuit and V_{CM} generator. **b** D flip-flops. **c** XYZ encoder. All transistor sizes (W/L) are in μm. Sizes without brackets are for a 65 nm technology and those within brackets are for a 0.13 μm one

where $V_{id} = V_{inp} - V_{inn}$ represents the differential input voltage. The two threshold[1] (switching) voltages of the quantizer ($\pm V_{TH}$) are the corresponding values of V_{id} when V_{op} or V_{on} reaches $(V_A - V_B)/2$, or when the differential output signal V_{od} $(= V_{op} - V_{on})$ is approximately $V_{DD}/2$, as shown in Fig. 4.2a.

The proposed 1.5-bit flash quantizer comprises another two inverters (INV_5 and INV_6), a current source (M_5) and two transistors operating in the triode region (M_{4A} and M_{4B}). The two extra inverters together with the D-type Flip-Flops (D-FFs) regenerate the output signals from INV_1 and INV_2, which strongly reduces the comparison (regeneration) time. Fig. 4.2b displays the simulated outputs of INV_1 and INV_2 as well as the outputs of INV_5 and INV_6 (V_{op} and V_{on}). The latter are connected to CMOS switched-inverter D-FFs (see Fig. 4.1b), followed by an XYZ encoder (Fig. 4.1c). The differential output signal, V_{od} (or V_{odL}, where L stands for latched), has three possible logic states −1, 0, and 1, as shown in Fig. 4.2a, which are used to build the outputs of the 1.5-bit flash quantizer. These possible logic

[1] Not to be confused with the threshold voltage of a transistor.

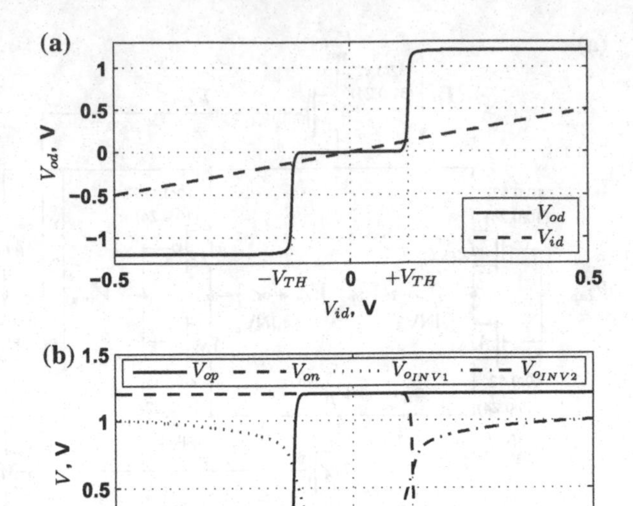

Fig. 4.2 Input-output waveforms: **a** differential output ($V_{od} = V_{op} - V_{on}$) and differential input signal, V_{id}. **b** Single-ended outputs of INV_1, INV_2, INV_5 and INV_6

states permit representing the outputs of the XYZ encoder, respectively given by X,Y, and Z. PMOS current source, M_5, has its gate voltage controlled by the output of the self-biasing inverter (INV_3), and is connected to the circuit in a negative feedback loop. This loop reduces the effect of supply and process variations. As an example, if V_{DD} increases, V_{SGP5} also increases producing an increase in the bias current, I_B. This will proportionally change the current in the main inverters. Since M_{P3} is tied to a constant gate and source voltage (V_{CM} and V_A), V_{SGP3} will remain constant but the rising current flowing through M_{P3} will cause V_C to increase. Therefore, $V_{SGP5} = V_{DD} - V_C$, will remain approximately constant and compensate the V_{DD} variation. Process variations have similar compensations through the negative feedback loop. Moreover, the feedback loop also stabilizes the bias voltages when variations in device parameters or operating conditions occur. Transistors M_{4A} and M_{4B} are always biased in the triode region, which is guaranteed by their high (and nearly constant) gate voltage, V_A. Since these two devices can have different dimensions, transistor M_{4B} introduces a degree of freedom, which can be used to optimize the temperature compensation, particularly in deep submicron CMOS technologies.

Comparing with the self-biasing amplifier/comparator in [10] the circuit proposed here has significant differences and improvements:

- two differential inverter comparator structures are merged by sharing the self-biasing inverter;
- two built-in threshold levels are obtained by suitable sizing of transistors M_{N1-3}, M_{P1-3};

- M_5 is biased in the saturation region rather than in the triode region, thus reducing the effect of supply voltage variations;
- the gate voltage of M_{4A} is V_A instead of V_C, which improves both stability and temperature compensation;
- as in [94, 139, 185, 186], two standard-sized inverters (INV_5 and INV_6) improve regeneration time and reduce metastability;
- transistor M_{4B} is added to further improve temperature compensation in deep-submicron CMOS technologies.

4.1.2 Circuit and Performance Analysis

4.1.2.1 Circuit Analysis

Advanced CMOS technologies lead to the reduction of the supply voltage, which in turn cause the transistors to operate in weak-to-moderate inversion regions. This situation becomes exacerbated in circuits with many stacked transistors, which is the case of the present circuit. Therefore, it is probable that all transistors of the three main inverters operate in weak-to-moderate inversion. The low biasing currents and with V_{OD} (overdrive voltage, $V_{OD} = V_{GS} - V_{TH}$) less than 100 mV [167] will lead to operation in these regions. The I_D-V_{GS} characteristic of these transistors is (approximately) exponential, given by

$$I_D = I_{D0}\frac{W}{L}e^{\frac{V_{GS}}{nV_t}} \tag{4.2}$$

where I_{D0} represents a characteristic current, W and L are the width and length of the transistor, n is the subthreshold slope coefficient, and $V_t = kT/q \approx 26\,\text{mV}$ (at room temperature) is the thermal voltage. Coefficient n depends on the biasing and thus is not accurately known. The values of n and I_{D0} can be extracted from simulation data. To extract the values of these parameters, the exponential approximation of the I_D-V_{GS} curve of M_{P2} depicted by the top plots of Fig. 4.3a, b is used. For the 0.13 µm technology, $n \approx 1.9$ and for the 65 nm technology, $n \approx 1.6$. As suggested in [167] a good procedure to verify a transistor in terms of its region of operation is to analyse its $\log(I_D)$-V_{GS} curve. This is depicted in the bottom plots of Fig. 4.3a, b for both technologies. As can be seen, the section where the graph is approximately linear defines the weak inversion region. Where the graph becomes more rounded defines the moderate inversion region. For the 0.13 µm process there is quite a reasonable approximation to the strong inversion equation near the switching V_{GS}, which does not occur for the 65 nm process, suggesting that, for this technology, the transistors operate in a weak-to-moderate inversion region.

To determine the switching threshold voltage, V_{TH}, a derivation from Eq. 4.1b and Eq. 4.2 is used. By writing Eq. 4.2 for $V_{GS}(I_D)$, and substituting it in Eq. 4.1b, we obtain

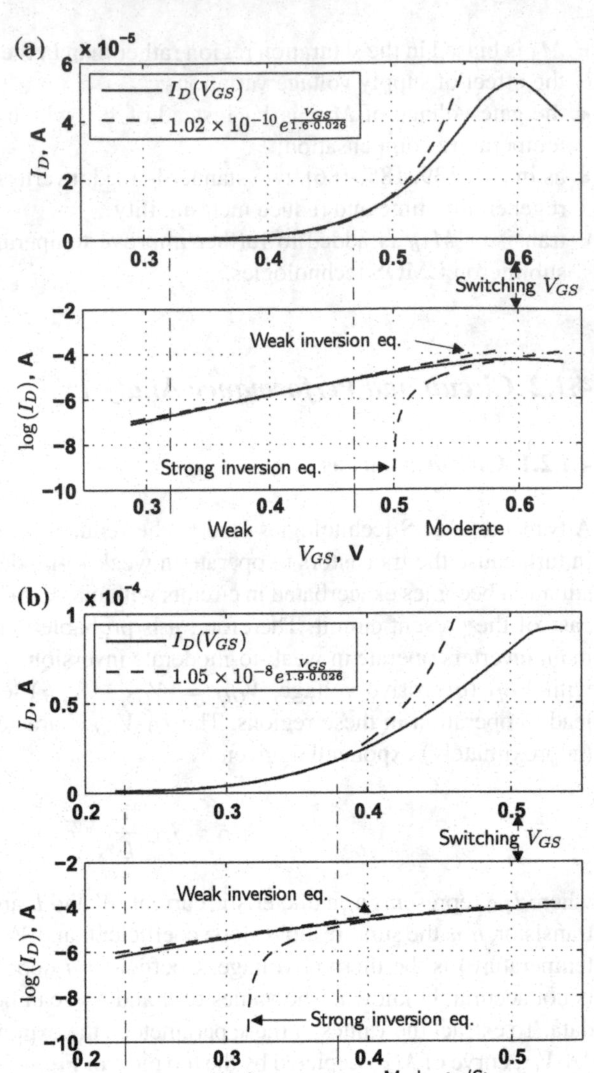

Fig. 4.3 $I_D - V_{GS}$ and $\log(I_D) - V_{GS}$ curves of the PMOS transistors of $\mathrm{INV}_{1,2}$ for: **a** 65 nm process and **b** 0.13 μm process

$$\begin{cases} -nV_t \ln\left(\dfrac{I_{DN2}}{I_{D0N2}}\dfrac{L_{N2}}{W_{N2}}\right) + nV_t \ln\left(\dfrac{I_{DN3}}{I_{D0N3}}\dfrac{L_{N3}}{W_{N3}}\right) - (V_{DS4A} - V_{DS4B}) = \dfrac{V_{TH}}{2} \\ +nV_t \ln\left(\dfrac{I_{DP2}}{I_{D0P2}}\dfrac{L_{P2}}{W_{P2}}\right) - nV_t \ln\left(\dfrac{I_{DP3}}{I_{D0P3}}\dfrac{L_{P3}}{W_{P3}}\right) - (V_{DS4A} - V_{DS4B}) = \dfrac{V_{TH}}{2}. \end{cases} \quad (4.3)$$

By taking into account that $I_{DN2} = I_{DP2}$ and $I_{DN3} = I_{DP3}$, by assuming that $I_{D0N2} = I_{D0P2}$, $I_{D0N3} = I_{D0P3}$, $L_{N2} = L_{P2}$, $L_{N3} = L_{P3}$, by summing both expressions of Eq. 4.3 and by applying simple logarithmic math, the following

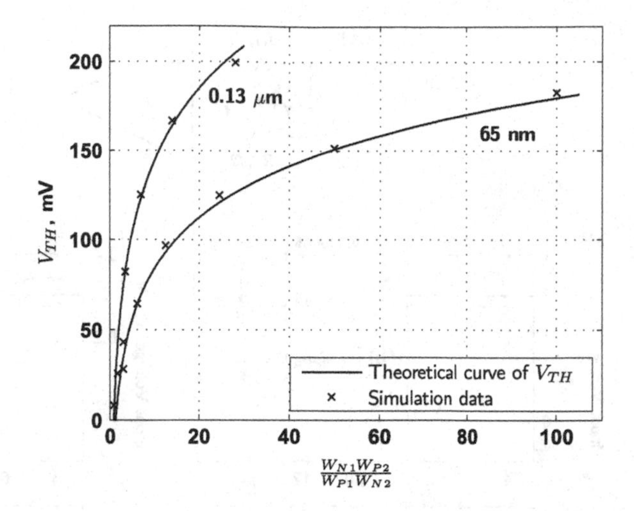

Fig. 4.4 Theoretical curve of the quantizer's threshold voltage, V_{TH}, and simulation data points, for the 65 nm and 0.13 μm technologies

expression for the switching threshold voltage is obtained

$$V_{TH} = FnV_t \ln\left(\frac{W_{N2}W_{P3}}{W_{P2}W_{N3}}\right) - (V_{DS4A} - V_{DS4B}), \tag{4.4}$$

where F is a fitting parameter. Equation 4.4 represents an expression that defines the switching threshold voltage of the 1.5-bit flash quantizer as a function of the transistors' widths (transistor lengths are easily included). Regarding $V_{DS4A,B}$, transistors $M_{4A,B}$ will operate in the linear region and therefore are easily sized using known equations. An expression similar to Eq. 4.3 and Eq. 4.4 can be derived combining Eqs. 4.1a and 4.2 for defining the negative threshold voltage, $-V_{TH}$. This is unnecessary because as long as $W_{N1} = W_{N2}$ and $W_{P1} = W_{P2}$, the threshold voltage will be symmetric around 0 V, i.e., $\pm V_{TH}$ will be obtained.

Figure 4.4 illustrates how the theoretical expression of Eq. 4.4 matches simulation data for several threshold voltages. This curve fitting allows the determination of the F parameter. This parameter is necessary to account for variations of n for different bias voltages and also to take into account that the transistors operate between weak and moderate inversion. The values for the fitting parameter are 1.2 and 1.0 for 0.13 μm and 65 nm technologies, respectively.

4.1.2.2 Kickback Noise

In latched comparators, kickback occurs during the switching of the positive feedback latch (PFBL). The large and rapid transients inject charge (kickback voltage) through parasitic capacitors into the input voltage, causing unwanted disturbances. In the proposed flash quantizer, when the D-FFs are clocked to produce outputs Q and $\overline{Q}$ (see Fig. 4.1b), the switch at the beginning of the D-FF is opened, which strongly

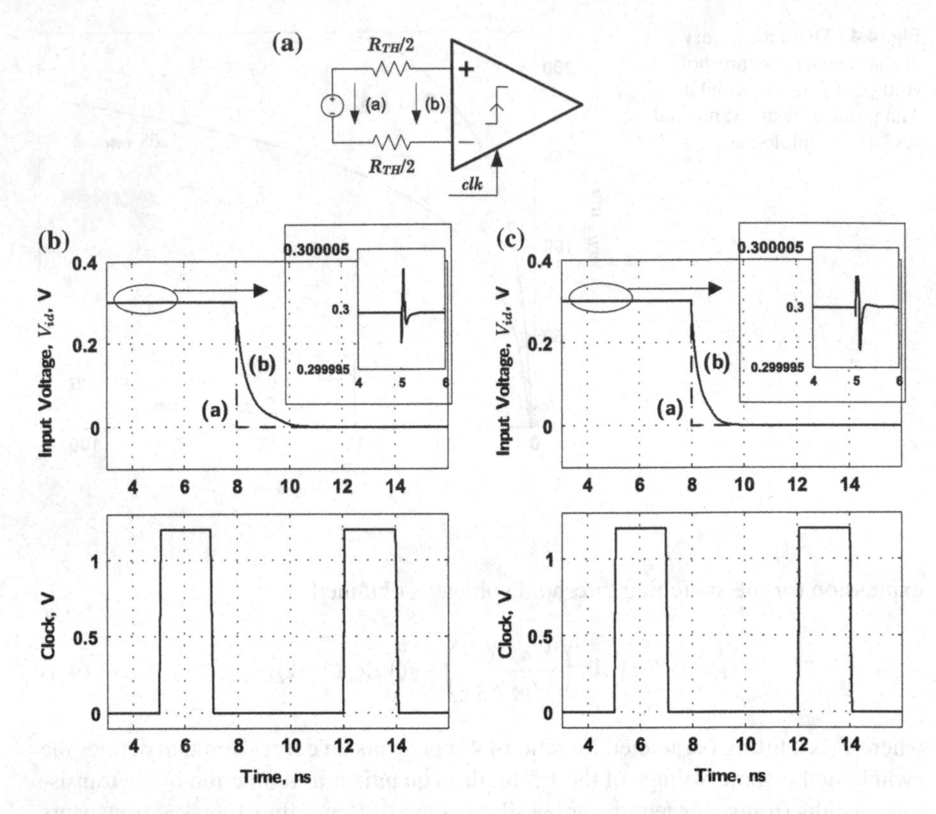

Fig. 4.5 Kickback noise analysis: **a** simulation setup. **b** Simulation results for 65 nm and **c** 0.13 μm technologies

reduces kickback. Besides this, from the output V_{od} to the input V_{id} there are a cascade of parasitic capacitors, i.e., series of the equivalent C_{gd} of $\mathrm{INV}_{5,6}$ with the equivalent C_{gd} of $\mathrm{INV}_{1,2}$, which reduces the parasitic capacitance from output to input. The kickback voltage is analysed using an identical simulation setup to the one proposed in [52], illustrated in Fig. 4.5a ($R_{TH} = 8\,\mathrm{k}\Omega$), with the results shown in Fig. 4.5b, c. The kickback during switching is approximately 4 and 3 μV, for the 65 nm and 0.13 μm technologies, respectively.

However, the proposed circuit's outputs, V_{od}, may vary before any clocking occurs, given its analog architecture. These outputs may change from rail-to-rail depending on the actual input. This will inject charge into the input voltage through the series parasitic capacitances mentioned above. Naturally the amount of charge injected is highly dependent on the sizing of the transistors of the inverters. For the 65 nm technology (transistor sizes are those shown in Fig. 4.1a) the injected charge results in an added voltage of 9.7 and 18.5 mV for a rising and falling output, respectively. For the 0.13 μm technology the added voltage due to charge injection is 6 and 12.6 mV for a rising and falling output, respectively. The proposed sampling circuit

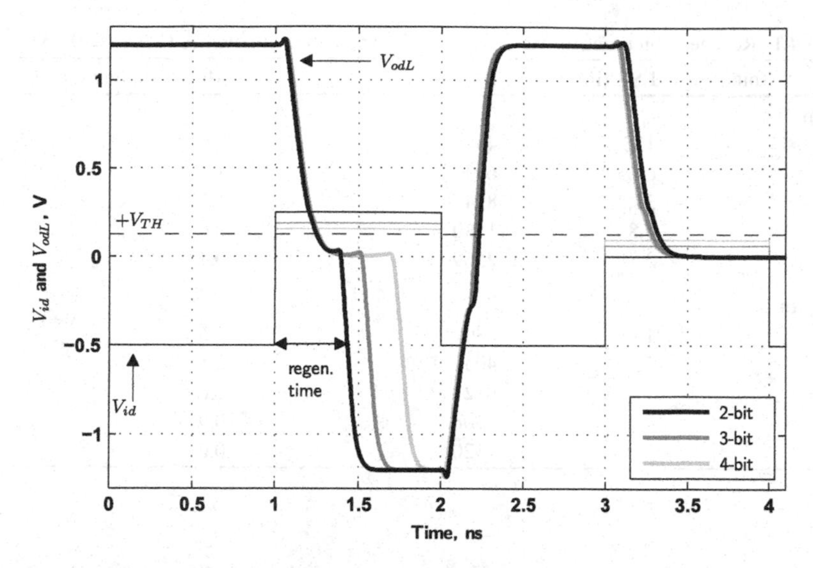

Fig. 4.6 Regeneration time of the output, V_{odL}, for various accuracies

(on the right of Fig. 4.1a) helps isolate the input voltage from any disturbances caused by kickback.

4.1.2.3 Regeneration Time

In the proposed architecture the regeneration time (or recovery time) may be defined as the time it takes the output, V_{odL}, to reach a valid logic value (considering the D-FFs are in a transparent state). It will be considered that 0.2 V represents a logic 0 and 1 V, a logic 1. To determine the regeneration time, the method suggested in [140] is employed. This method assesses the flash quantizer's speed performance by testing it for two extreme cases. The first case consists of a full-scale input, V_{FS}, that changes to $V_{TH} + 1\,\text{LSB}$. For the second case, the input goes from V_{FS} to $V_{TH} - 1\,\text{LSB}$ with the objective of causing indecision, and therefore the output needs to recover. The time it takes the outputs to reach logic values determines the maximum operating rate of the flash quantizer.

As can be seen by the voltage waveforms of Fig. 4.2, for each threshold voltage, one of the outputs, V_{op} or V_{on}, changes polarity. Depending on the input voltage, both outputs might end up changing, e.g., when the input changes from $+V_{FS}$ to below $-V_{TH}$ or from $-V_{FS}$ to above $+V_{TH}$. This is considered the worst-case, therefore, the regeneration time is simulated for this situation. The resultant waveforms (V_{id} and V_{odL}) for three accuracy situations (2, 3, and 4-bit) are illustrated in Fig. 4.6, for the 65 nm case only. The input goes from $-V_{FS}$ to $+V_{TH} \pm 1\,\text{LSB}$, with $V_{FS} = 0.5\,\text{V}$, $V_{TH} = 0.125\,\text{V}$, and $V_{REF} = 0.5\,\text{V}$. For the first test, $+V_{TH}+1\,\text{LSB}$, it can be seen that the output has an initial change of polarity, due to the negative threshold voltage, and

Table 4.1 Regeneration times for various accuracies and both technologies ($V_{REF} = 0.5$ V)

Accuracy (bits)	LSB (mV)	Regeneration time (ps)	Max. sampling rate (GS/s)
65 nm			
2	125	484	2.0
3	62.5	615	1.6
4	31.25	814	1.2
6	7.8	1580	0.6
8	2	2569	0.4
0.13 μm			
2	125	280	3.5
3	62.5	405	2.5
4	31.25	612	1.6
6	7.8	1108	0.9
8	2	1520	0.65

then another change in polarity, much more prolonged due to the positive threshold. The second change is more prolonged because the input voltage is very close to the threshold voltage (+1 LSB). The closer the input is to the threshold voltage, the longer the regeneration time.[2] The output then returns to 1.2 V because the input is at $-V_{FS}$ again and this starts the second test: the input goes from $-V_{FS}$ to $+V_{TH} - 1$ LSB. As can be seen, there is the initial change of polarity (due to $-V_{TH}$), but, although the input is close to the threshold voltage, the output stays constant and no indecision is visible.

Table 4.1 summarizes the regeneration times for various accuracies and for both technologies. Note that the flash quantizer sized for each technology was not optimized for speed. Therefore, the regeneration times for the 0.13 μm technology appeared faster. Note also that, the current consumption for the 65 nm case is less than half that of 0.13 μm case (see Table 4.3 for details).

4.1.2.4 Probability of Metastability

The proposed quantizer relies on a pipeline architecture as shown in Fig. 4.7, i.e., a cascade of open-loop amplifiers (inverters), to produce the final outputs. Therefore, assuming that the D-FFs are in writing mode there is, in fact, a cascade of four inverters before the XYZ encoder. These inverters are composed of INV_1 (or INV_2) followed by INV_5 (or INV_6), and two inverters from the D-FF (see Fig. 4.1). In order to calculate the final time constant of this kind of architecture, the method for multistage amplifiers proposed in [177] is used.

[2] With an input exactly equal to the threshold voltage, the regeneration time would be infinitely long, leading to a situation of metastability.

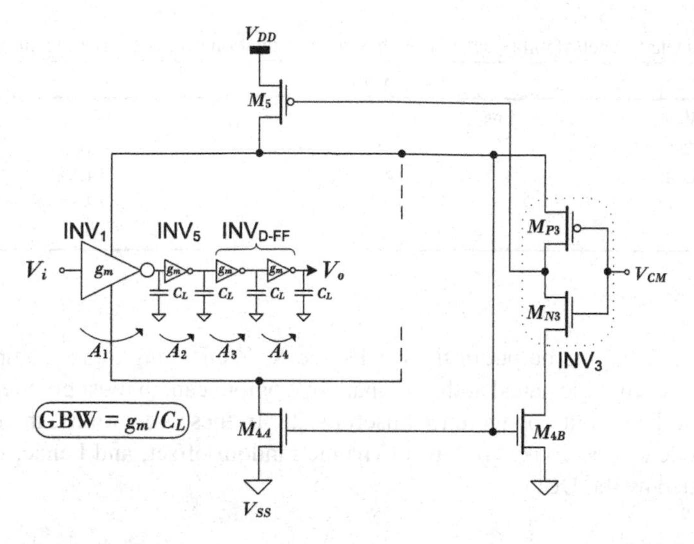

Fig. 4.7 Equivalent circuit for metastability analysis considering a multistage approach. Single-ended shown for simplicity

The time constant is defined by

$$\tau = \tau_m \left(\frac{V_o}{V_i} \times N! \right)^{1/N} = \frac{1}{\text{GBW}} \left(\frac{V_o}{V_i} \times N! \right)^{1/N}, \quad \text{GBW} = g_m / C_L, \tag{4.5}$$

where τ_m is given by the inverse of the averaged GBW of the four cascaded inverters,[3] N represents the number of cascaded inverters in the signal path, and V_o and V_i are the output and input voltages respectively. Using the same process to obtain the probability of metastability as shown in [191], it follows that

$$P_E = 2 \frac{V_{oe}}{V_{FS}} \frac{N!}{(t_r \text{GBW})^N}, \tag{4.6}$$

where V_{oe} indicates the output voltage range that will cause a metastability error, V_{FS} is the full-scale range of the input signal, and t_r is the available time for comparison.

Assuming the values given in Table 4.2 and substituting them in Eq. 4.6, the probability of metastability can be shown to be around 10^{-7} for both technologies (the transistor sizes are those shown in Fig. 4.1a). These probability values can be reduced by increasing the number of pipelined inverters, as well as optimizing the sizing of the inverters for maximum GBW. However, by increasing the number of cascaded inverters, there will be a sacrifice of power, area, and latency. Notice that in pipeline, algorithmic or multi-step flash ADCs, the effect of metastability is in part relaxed due to the digital correction logic (DCL) scheme. Although a valid logic level might

[3] The averaged GBW is used here as an approximation, to simplify the calculations.

Table 4.2 Data for metastability calculation based on simulation results for both technologies

Parameters	65 nm	0.13 μm
$V_{oe} = 0.2\,\text{V}$, $V_{FS} = 1\,\text{V}$, $t_r = 1\,\text{ns}$, $N = 4$		
GBW_1 (GHz)	3.9	4.3
GBW_{2-4} (GHz)	14.9	14.5
$\overline{\text{GBW}}$ (GHz)	12.1	12
P_E	10^{-7}	10^{-7}

not be provided at the output of the D-FFs, the XYZ encoding logic (composed of various cascaded logic gates) at the comparator's output can, in most practical cases, reconstruct the output voltage into a useful state. In these conditions, the effect in the sub-code can be considered as a dynamic random offset, and hence, it is also compensated by the DCL.

4.1.2.5 Offset

One the main limitations of comparator circuits is offset, i.e., deviations to the nominal switching threshold. These deviations can limit the performance of ADCs, namely by reducing or increasing the ADC's full-scale range. Offsets may occur due to mismatch between symmetric devices or in components that alter the operating point of the circuit. In the proposed flash quantizer, symmetric transistors are M_{P1} and M_{P2}, and M_{N1} and M_{N2}. Any mismatch between these devices will offset the threshold voltage. Regarding devices that may alter the circuit's operating point, these are all the others. Of these devices the ones that do not have a large influence on the threshold voltage are M_5, M_{4A}, and M_{4B}, while transistors M_{P3} and M_{N3} have a significant effect.

Mismatch may exist in various transistor parameters, namely, V_{TH}, W, L, C_{ox}, and μ. Mathematically, these mismatches are given by [93,131]

$$\Delta V_{TH} = \frac{A_{V_{TH,k}}}{\sqrt{WL}} \quad k = \{N, P\}, \tag{4.7a}$$

$$\Delta\beta = \frac{A_{\beta,k}}{\sqrt{WL}} \quad k = \{N, P\}, \tag{4.7b}$$

where $\beta = \mu C_{ox} W/L$, and $A_{V_{TH}}$ and A_β are area proportionality constants defined by the process technology. It has been demonstrated in [92,141], that the total offset referred to a transistor's gate is given by

$$V_{os,gate} = \frac{(V_{GS} - V_{TH})}{2}\left(\frac{\Delta(W/L)}{W/L}\right) + \Delta V_{TH}. \tag{4.8}$$

To determine the total input-referred offset voltage, $V_{os,in}$, a set of transfer functions (H_i) need to be determined, such as the H from each offset voltage source ($V_{os,i}$) to the output, and the open-loop DC gain ($A = V_{od}/V_{id}$) of the flash quantizer. Each offset voltage is referred to the input by multiplying the respective $V_{os,i}$ by the square of its transfer function and then dividing the result by the DC gain squared. It is important to remember to take the square (variance) of Eq. 4.8. The final input-referred offset voltage is then the root of the sum of each input-referred V_{os}. Considering only the ΔV_{TH} mismatch of Eq. 4.8, which is the largest contributor (this simplification still yields a good approximation), the input-referred offset is given by,

$$V_{os,in} = \sqrt{\sum_{i=1}^{n} \left(\frac{A_{VTH,i}^2}{(WL)_i} \frac{H_i^2}{A_{DC}^2} \right)}. \tag{4.9}$$

where n represents each transistor.

The transfer functions and open-loop DC gain are too complex to maintain a sufficient level of accuracy to gain insight to the circuit and, therefore, are not shown here. Nevertheless, a method has been given to determine the offset voltage and simulation results will be shown instead. Monte Carlo simulation results (500 cases) for both technologies are shown in Fig. 4.8, for the negative (top row) and positive (bottom row) threshold voltages. The graphs on the left of Fig. 4.8 correspond to the worst-case threshold voltage that occurs below the nominal V_{TH} ($\pm$125 mV), while the graphs on the right are for the worst-case threshold above the nominal V_{TH}. For these simulations the transistor sizes of Fig. 4.1a are used. Besides mismatch variations, process, supply voltage and temperature (PVT) are also varied. The variations considered are typical-typical (tt), slow-slow (ss), and fast-fast (ff) for process, 1.2 V $\pm$ 5 % for the supply voltage, and −40, 27, and 85 °C for temperature. For the 65 nm technology, with nominal temperature (27 °C) and supply voltage (1.2 V), $\pm V_{TH}$ has an average value of 125 mV and a standard deviation of 8.1 mV. The minimum threshold voltage $V_{TH_{min}} = \pm 116.2$ mV (in average) is obtained for −40 °C and 1.26 V supply. The maximum value, $V_{TH_{max}} = \pm 128.9$ mV, is obtained for 27 °C and 1.14 V supply. The worst-case standard deviation obtained is 8.5 mV. Regarding the 0.13 μm technology, in nominal conditions V_{TH} has an average value of approximately 125 mV and a standard deviation of 7 mV. The minimum threshold voltage $V_{TH_{min}} = \pm 109$ mV is obtained for 85 °C and 1.26 V supply. The maximum value, $V_{TH_{max}} = \pm 129$ mV, is obtained for 27 °C and 1.14 V supply. The worst-case standard deviation obtained is 7.7 mV.

Figure 4.9 depicts the threshold voltage (average value) deviation with respect to PVT variations. The objective of this graph is to illustrate the tendency of the threshold voltage as a function of the aforementioned PVT variations. For the 65 nm technology, V_{TH} varies approximately 6 % for the full supply voltage range (considering all process and temperature variations). For a given supply voltage, V_{TH} varies with temperature approximately 6.5 % and for the three process corners considered, less than 2.5 %. Regarding the 0.13 μm technology, V_{TH} varies approximately 13 % for the full supply voltage range and for all process and temperature variations. For a specific voltage, V_{TH} varies with temperature approximately 8.5 % and with process corners less than 12 %. Figure 4.9 and the described results indicate that the 65 nm technology shows much improved results, specially concerning process and supply voltage variations. Furthermore, by analysing the graphs of Fig. 4.9, which have an identical y-axis limits for comparison, it can be seen that the range of the results for the 65 nm technology is smaller.

4.1.2.6 Common-Mode Sensitivity

Figure 4.10 depicts the threshold deviation due to variations in the common-mode (CM) voltages of the circuit. The results indicate that V_{TH} deviates ±2 and ±7 % for the 65 nm and 0.13 μm nodes, respectively. The simulation considered a ±25 mV variation in V_{CM}. Therefore, the circuit exhibits low sensitivity to variations in the CM voltages.

The above simulation analysis assumed that both the CM voltage of the input signal and the voltage at the input of INV_3 are the same. However, simulations show that the latter can be used as a control voltage, to correct for deviations in the threshold voltages. With an adaptive biasing circuit capable of generating this V_{CM} dependent on PVT conditions, the threshold voltages may be kept more constant, therefore reducing the offset. Moreover, depending on the sizing and biasing conditions of the circuit, this V_{CM} may even be used to generate more than one threshold voltage.

4.1.2.7 Proof-of-Functionality

As a proof-of-functionality, the proposed flash quantizer circuit is employed in all sub-ADCs of a real 7-bit 500 MS/s pipeline converter, designed in the 0.13 μm technology. The converter is composed of five 1.5-bit stages and a final 2-bit stage. All 1.5-bit sub-ADCs rely on the circuit shown in Fig. 4.1. For the final 2-bit stage, the quantizer's thresholds are ±250 mV and 0 V. For the ±250 mV thresholds the proposed circuit architecture is used, but with a resizing of the transistors' dimensions. The 0 V threshold is easily achieved by eliminating INV_1 (or INV_2) and by making INV_3 equal, i.e., the dimensions of the transistors, to the inverter that remained (INV_1 or INV_2).

Figure 4.11 shows the electrically simulated 8192-point FFT, DNL and INL results of the pipeline ADC for a 200 MHz input signal. A SFDR of 45.5 dB and an SNDR of 41.4 dB is obtained which leads to an ENOB of 6.6 dB. Concerning DNL and INL, the maximum absolute values obtained are −0.32 and −0.59 LSBs respectively. These results demonstrate the practical usefulness of the proposed 1.5-bit flash quantizer circuit since good DNL, INL, and SNDR performance are achieved.

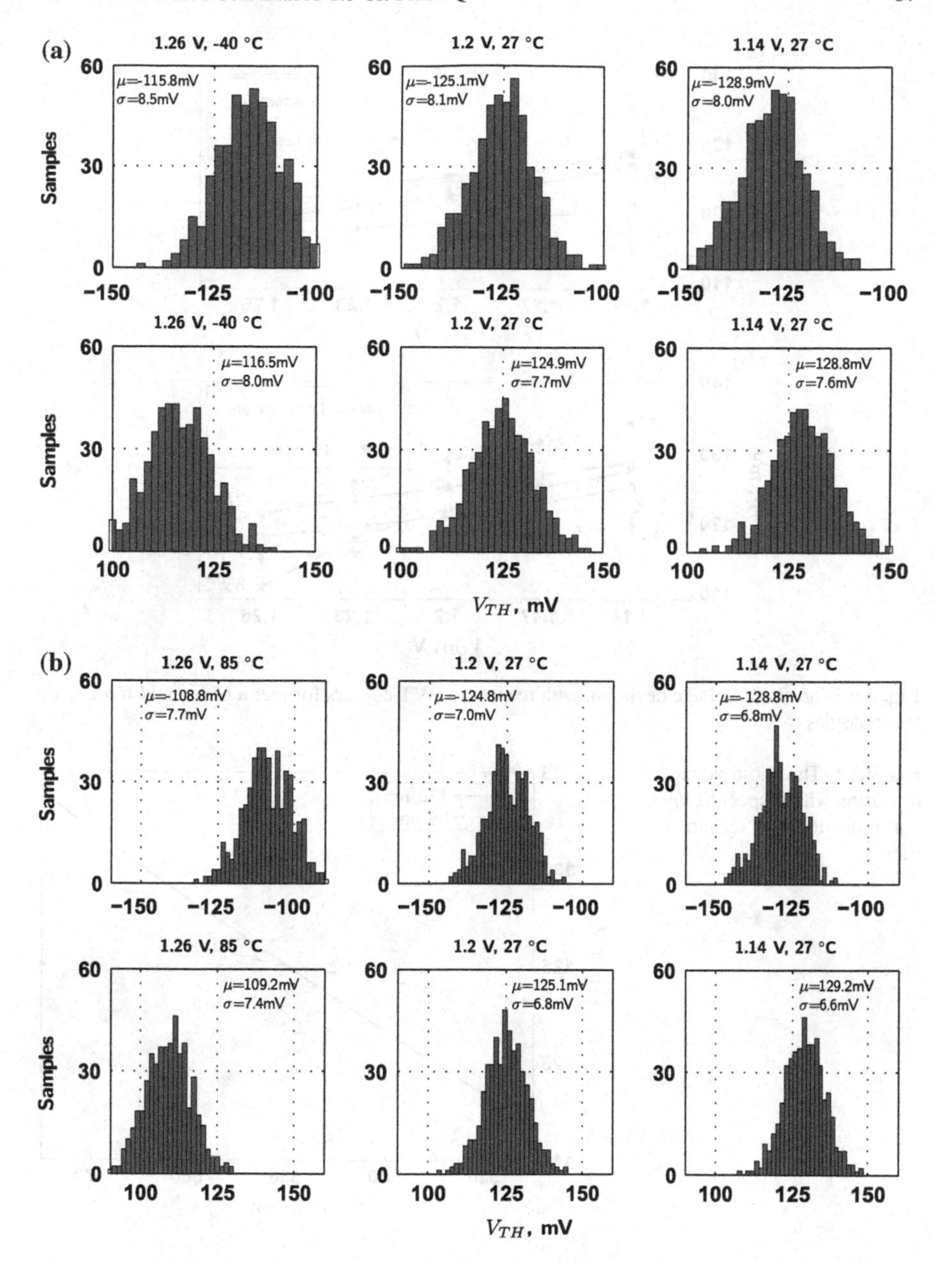

Fig. 4.8 Histograms for 500 Monte Carlo simulations considering PVT corners and mismatch for the: **a** 65 nm process and (**b**) 0.13 μm technology. *Top* and *bottom row* show the negative and positive threshold voltage, respectively. The mean value and standard deviation of the data are represented by μ and σ, respectively

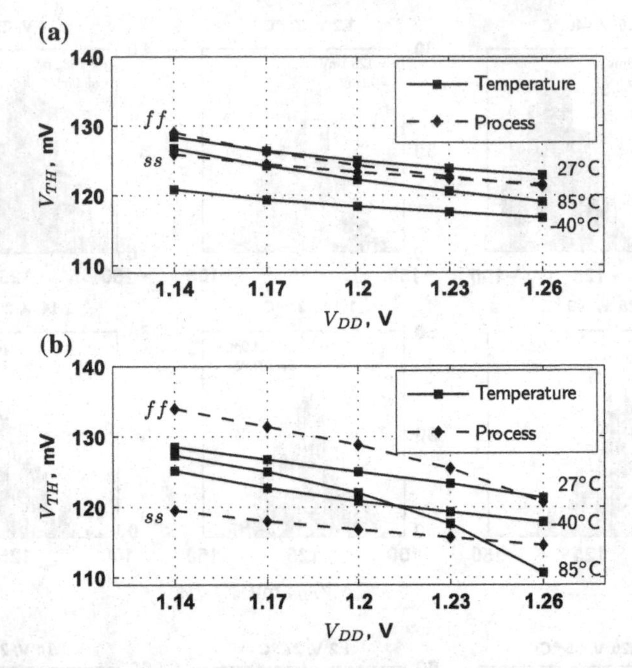

Fig. 4.9 Threshold voltage deviation with respect to PVT corners for the: **a** 65 nm and **b** 0.13 μm technologies

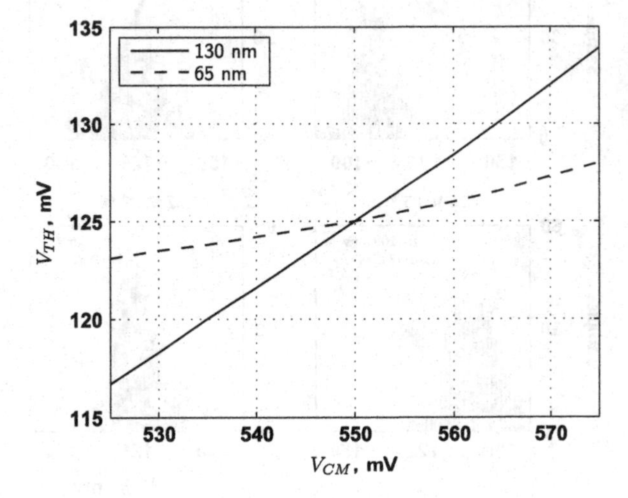

Fig. 4.10 Threshold voltage deviation with respect to V_{CM} variations for both technologies

4.1.3 Design Procedure

To achieve a specific V_{TH}, the following steps may be considered as a design procedure and guidelines to an approximately accurate threshold voltage, when taking into account PVT variations.

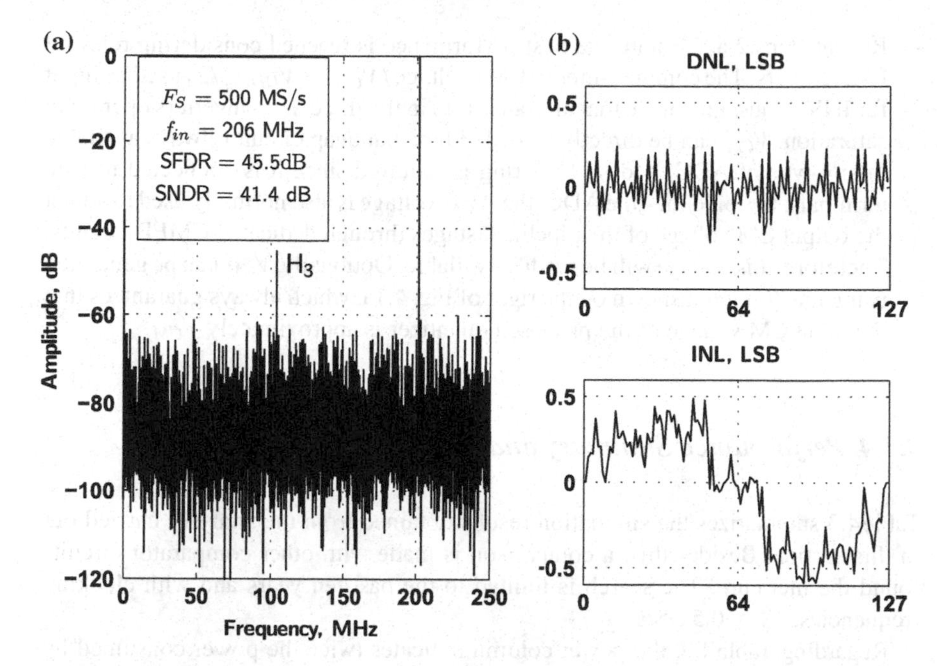

Fig. 4.11 Proof-of-functionality of a 7-bit 500 MS/s pipeline ADC employing the proposed quantizer in all pipelined stages: **a** 8192-point FFT. **b** DNL and INL

1. Using Eq. 4.4, the widths (W_i) of the transistors of INV_1, INV_2 and INV_3, are obtained. Current source, M_5, is sized for an average current of the order of 100 μA (or even less, depending on the power budget and desired speed) and with an overdrive, V_{OD}, of about 100 mV (biased in moderate inversion). Regarding V_A, it is hard to define this voltage since M_5 is a current source. However boundaries can be defined: V_A should be lower than $V_{DD} - V_{OD} = 1.1$ V, otherwise transistor M_5 may leave the saturation region and, on the other hand, V_A should be set above 1 V for $\text{INV}_{1,2}$ and INV_3 to operate correctly. For a current of 100 μA, the NMOS resistors (M_{4A} and M_{4B}) are sized to have a voltage drop of about 50 mV. To avoid undesired short-channel effects and offsets due to process variations, channel lengths (L_i) should be above the minimum. To limit offsets due to random transistor mismatches, minimum device areas ($W_i L_i$) should be set well above 4 μm^2, particularly for the transistors of the inverters.
2. Transient simulations are performed with slow ramp signals as a differential input to verify the values of $\pm V_{TH}$ for the typical case of parameter values (tt, 1.2 V, and 27 °C). This simulation is repeated for all main PVT corners (T_{min} and T_{max}, VDD_{min} and VDD_{max}, and process corners ss and ff).
3. Compensation over the entire temperature range is optimized, by adjusting M_{4A} and M_{4B} with different dimensions. Since this will affect V_{TH}, the width of M_{N1} and M_{N2} should be fine tuned to restore the value of V_{TH}.

4. Repeat steps 2 and 3 until the best performance is reached considering relevant PVT corners. The common-mode (CM) voltage ($V_{CM} \approx V_{DD_{min}}/2$) used as input for INV$_3$, must guarantee that all transistors in the three main inverters operate in saturation. V_{CM} can be directly provided by a bandgap circuit (always available in any type of ADC) and no buffering is required since it is connected to gate terminals. In a pipeline-like ADC, this V_{CM} voltage is also normally used to adjust the output CM voltage of the pipelined stages through dedicated CMFB circuits. Therefore, this voltage will be readily available. Optionally, V_{CM} can be generated using the SC circuit shown on the right of Fig. 4.1a, which always guarantees that the input CM voltage of the proposed quantizer is approximately V_{CM}.

4.1.4 Performance Summary and Comparison

Table 4.3 summarizes the simulation results obtained from the analyses carried out in this section. Besides this, a comparison is made with other comparator circuits found the literature. The search is limited to the past ten years and with clocking frequencies, $F_S \geq 0.5$ GS/s.

Regarding Table 4.3, the power column indicates twice the power consumed by the respective comparator. This is done for a fair comparison, since the proposed work is a 1.5-bit flash quantizer, i.e., it has two comparators merged into one circuit. The resolution column indicates the accuracy of the voltage step used to obtain either the metastability or the regeneration (comparison) time. Some authors indicate the minimum voltage to obtain a given metastability, while others indicate the step voltage (or differential input) used for the regeneration time analysis. The maximum accuracy (N) possible, for each of the presented circuits, is related with the PVTM-3σ offset (Off$_{3\sigma}$), given by $N < \log_2(V_{FS}/\text{Off}_{3\sigma})$, where V_{FS} is the differential full-scale voltage of the input signal.

The results of Table 4.3 indicate that the proposed quantizer has a relatively slow regeneration time, due to the fact that most of the other comparators employ a positive feedback latch, which accelerates their decision time. The proposed quantizer achieves an average metastability probability error, a better than average PVTM-3σ offset and one of the leading energy efficiencies among the state-of-the-art. Notice however, that the results of all (except [122]) comparators are from measured data, whereas the results of the proposed circuit are extracted from electrical simulations. Regarding the applications envisaged by each comparator, most may be applied in A/D converters, while others are designed for receiver front-ends [121], hence its high clocking frequency.

Table 4.3 Simulated performance summary and comparison with other comparator circuits found in the literature in the past decade for clocking frequencies, $F_S \geq 0.5$ GS/s

	Tech. (μm)	Supply voltage (V)	Power[a] (μW)	F_S (GS/s)	Resolution (bits)	Regeneration Time (ps)	Metastability BER[b]	Offset PVTM-3σ (mV)	EE[c] (fF/conv.-step)
This work	0.065	1.2	133	2	2	484	10^{-9}	27	33
This work	0.13	1.2	231	2	2	280	10^{-10}	33	58
[112]	0.09	1	80	1	10	122	–	5.1[d]	40
[111]	0.09	1.2	78	0.5	–	–	–	11.4[d]	78
[148]	0.09	1.2	450	2	~2	~60	–	24	113
[57]	0.12	1.5	720	2	6	–	10^{-9}	30	180
[61]	0.065	1.2	2600	7	~5	64	–	33	186
[60]	0.12	1.5	1168	3	~7	–	10^{-9}	48.3	195
[58]	0.12	1.5	1624	4	2	–	10^{-9}	150	203
[175]	0.18	1.8	700	1.4	–	–	–	13.1	250
[122]	0.13	1.2	284	0.5	2	165	–	60.9	284
[59]	0.12	1.5	5300	6	~2	–	10^{-6}	69	442
[128]	0.18	1.8	3700	3.84	10	250	10^{-13}	44.4	482
[121]	0.11	1.2	10080	10	–	–	10^{-12}	–	504
[62]	0.065	1.2	5760	5	6	104	10^{-9}	6	576
[153]	0.35	3.3	4000	1	6	–	–	150	2000
[176]	0.35	3.3	6600	1.2	6	–	–	0.6[d]	2750

[a] For the presented references, this column indicates twice the power consumed by the respective comparator [b] Bit Error Rate [c] Energy Efficiency, $EE = \text{Power}/2^{\text{ENOB}} F_S$, where ENOB = 1-bit [d] Requires calibration

4.2 Two-Stage Inverter-Based Self-Biased Opamp

The power reduction necessity and low supply voltage tendency of modern CMOS technologies has driven the evermore challenging design of amplifiers to have multiple gain stages and possibly an output driver stage. In single stage amplifiers, high DC gain is achieved using cascode devices, but this leads to reduced output swing (OS) due to the low supply voltage. Maintaining a high DC gain requirement but now demanding high OS would naturally lead to the use of a two-stage amplifier (probably fully differential to double the OS). In two-stage amplifiers, where compensation is inevitable, the gain-bandwidth product, GBW (usually a specification: achieve highest possible GBW), is given by $\text{GBW}(I) \approx g_{m1}(I)/C_C$, where g_{m1} represents the transconductance of the input differential pair, C_C, the compensation capacitance, and I, the biasing current of the input stage. Admitting that the opamp is biased with constant current, I, and with C_C being mainly imposed by thermal noise constraints, then, one way of achieving higher GBW relies on using an inverter structure in the input stage. This type of input stage, if properly sized, effectively doubles the transconductance. Now $g_{m1} = g_{mP} + g_{mN}$, where g_{mP} and g_{mN} respectively represent the transconductance of the PMOS and NMOS transistors that constitute the inverter structure [92]. In the literature it is possible to find numerous other techniques for increasing and compensating the GBW of single- and multi-stage amplifiers [43, 63, 96, 145], among others, mostly relying on multipath and feedforward techniques.

Single-stage inverter input amplifiers have already been presented in the past [10, 174, 180]. The inverter amplifier of [10] adds an interesting feature which is self-biasing. The advantages of amplifier self-biasing techniques are well known [10,106]:

- they simplify the implementation of amplifiers by removing biasing circuitry, thus saving power and die area;
- any mismatches and variations in circuit performance due to deviations in biasing voltages are highly attenuated;
- they enable circuits to be more insensitive to PVT variations.

In amplifiers, self-biasing techniques have mainly been applied to complementary folded-cascode and single stage inverter-based topologies [10, 106, 137].

The work developed in this book proposes combining self-biasing and inverter-input structures in a two-stage amplifier enhancing its efficiency and achieving higher PVT robustness.

4.2.1 Principle of Operation

The proposed two-stage inverter-based self-biased amplifier is depicted in Fig. 4.12. It consists of two cascaded inverter stages with approximately the same topology.

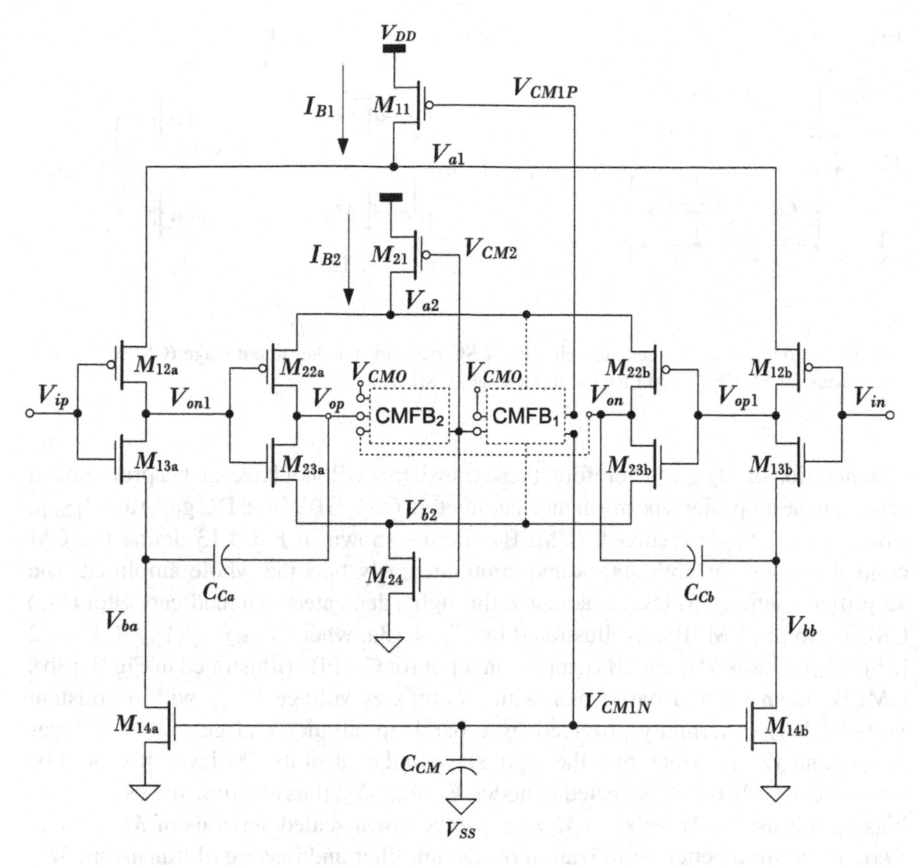

Fig. 4.12 Proposed two-stage inverter-based self-biased amplifier

The input stage consists of an inverter input pair ($M_{12a,b}$ and $M_{13a,b}$) connected to current sources, M_{11} and M_{14}. The latter devices bias and control the common-mode (CM) level of the input stage. The output stage has a similar topology (except that nodes V_{ba} and V_{bb} are connected together to form node V_{b2}) and its input pair consists of transistors $M_{22a,b}$ and $M_{23a,b}$. The biasing of the output stage and its output CM level are controlled by M_{21} and M_{24}. Given M_{21} and M_{24}'s biasing voltage, V_{CM2}, these devices will probably operate in the triode/saturation boundary region [10]. Regarding compensation, node V_{ba} and V_{bb} have been separated, for the connection of the compensation capacitors, $C_{Ca,b}$, thus avoiding the inefficient conventional Miller compensation [5,64]. In a preliminary analysis, this compensation can be described as a cascoded-Miller type, in the sense that $C_{Ca,b}$ are connected to low impedance source nodes. This method is adequate because there is no direct forward-path between the input and output of the output stage, thus avoiding the presence of a possible positive low-frequency zero in the transfer function (TF). Given that this method of compensation was used (to avoid Miller compensation), there is source

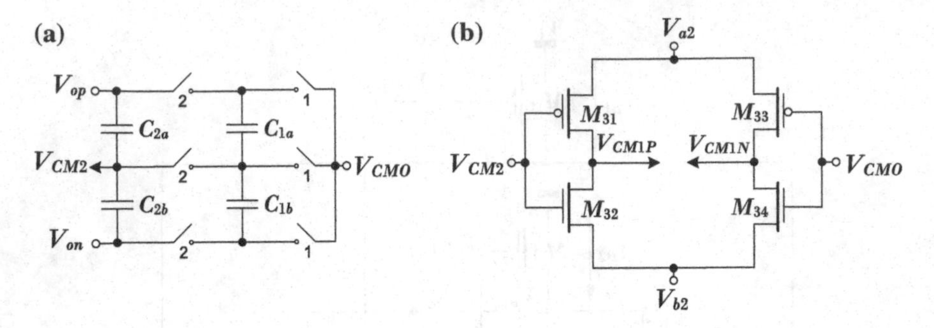

Fig. 4.13 Common-mode feedback circuits: **a** SC network for the output stage ($CMFB_2$) and **b** continuous time CMFB circuit for the input stage ($CMFB_1$)

degeneration of $M_{13a,b}$. Therefore, these transistors will not have such a pronounced effect on the amplifier's performance parameters (e.g., GBW and DC gain) as $M_{12a,b}$. The common-mode feedback (CMFB) circuits shown in Fig. 4.13 define the CM control voltages of both stages and simultaneously bias the whole amplifier. The amplifier's output CM level is adjusted through a dedicated switched-capacitor (SC) CMFB circuit, $CMFB_2$, as illustrated by Fig. 4.13a, where $V_{CM2} = (V_{op} + V_{on})/2$ [35]. V_{CM2} biases M_{21} and M_{24}, and is an input for $CMFB_1$ (illustrated in Fig. 4.13b). $CMFB_1$ is an inverter-based pair which compares voltage V_{CM2} with a constant voltage, V_{CMO} (normally provided by a bandgap circuit), and generates voltages V_{CM1P} and V_{CM1N} which bias the input stage and control its CM level. It should be noticed that $CMFB_1$ is connected to nodes V_{a2} and V_{b2}, thus avoiding the use of extra biasing transistors. Transistors M_{31-34} can be down-scaled versions of $M_{22a,b}$ and $M_{23a,b}$, but for a better optimization of the amplifier and biasing of transistors M_{11} and $M_{14a,b}$, these should be sized separately. Capacitor C_{CM} (Fig. 4.12) is placed at node V_{iCM1N} for CM stabilization.

Self-biasing voltages V_{CM1P}, V_{CM1N} and V_{CM2} are connected to the main amplifier through feedback loops thus reducing the effects of PVT variations and the effects caused by differential-mode (DM) and CM input variations. As an example of compensation: suppose the voltages on nodes V_{op} and V_{on} have already settled, if V_{DD} increases, the source-gate voltage of M_{21}, V_{SG21}, also increases producing an increase in the bias current I_{B2}. This increase will change proportionally the current in the two output inverters, increasing the output CM voltage. As a consequence, $CMFB_2$ will produce a higher V_{CM2} output control voltage forcing V_{SG21} and I_{B2} to their initial value, thus compensating the V_{DD} variation. A similar analysis may be carried out for the biasing and CM control of the input stage. For example, if the input CM level rises, the input stage's CM output falls causing the output stage's CM level to rise. V_{CM2} follows this rise, and makes V_{CM1P} fall and V_{CM1N} rise. As V_{CM1P} falls, V_{SG11} rises, forcing more current into the input stage's inverters, forcing its output to rise, thus, opposing its initial tendency. However, the opposite occurs with V_{CM1N}, which reinforces the output's initial tendency. Consequently, there is a negative feedback loop between V_{CM1P} and $V_{OCM1} (= (V_{op1} + V_{on1})/2)$, but, there is a positive feedback loop between V_{CMIN} and V_{OCM1}. The amount of variation of the aforementioned CM signals is dependent on the transistor sizing, but they are expected to be small, given the large CM input range of the amplifier and of CMFB_1 (due to the use of inverter structures), and due to the adopted self-biasing scheme.

Process and temperature variations have similar compensations through the negative feedback loops. Another fact that supports the good compensation for PVT variations is the completely complementary (half PMOS and half NMOS) design of the circuit.

4.2.2 Circuit Analysis

4.2.2.1 Differential-Mode Analysis

The small-signal DM equivalent circuit is shown in Fig. 4.14, for simplicity only half the circuit is shown and C_L represents a capacitive load. The DM TF can be extracted by applying Kirchhoff's laws, or by using a symbolic analyser, such as MATLAB [108]. The TF with no simplifications and no assumptions is too complex to represent and obtain a straightforward analysis. Considering some simplifications (namely, nulling parasitic capacitances C_{sb} and C_{db}) and tolerating errors up to 25 % (nulling smaller terms), a simplified TF is given by

$$\frac{v_o}{v_i}(s) = \frac{\begin{array}{r}[(C_{gs13}+C_{gd2})C_{gd1}+C_{gs13}C_{gs2}]C_C s^3 + [(C_{gd1}+C_{gs2})g_{m13}-C_{gd1}g_{m2}]C_C s^2\\ +(g_{m13}g_{m2}+g_{m12}g_{m23})C_C s\\ +g_{m12}g_{m13}g_{m2}\end{array}}{\begin{array}{l}(C_{gd1}+C_{gd2}+C_{gs2})C_C C_L s^3\\ \quad + [C_C C_{gs2} g_{m13} + ((C_{gd1}+C_{gd2}+C_{gs2})g_{m13}+C_{gs2}g_{mb13})C_L]s^2\\ \quad + (g_{mb13}g_{m23}+g_{m13}g_{m2})C_C s + (g_{m13}g_{o12}+g_{o13}g_{o14})g_{o2}.\end{array}} \tag{4.10}$$

where g_{mij} and g_{oij} represent the transconductance and the output conductance of M_{ij}, respectively, and g_{mb13} is M_{13}'s bulk transconductance, caused by bulk effect. C_{gdij} and C_{gsij} represent the gate-drain and gate-source parasitic capacitance, and C_C the compensation capacitance. For the sake of simplicity, minor textual generalizations are used and can be understood through the following example: $g_{ox} = g_{ox2} + g_{ox3}$, $x = \{1, 2\}$. This is valid for all parasitic capacitances, output conductances, and transconductances. The extracted TF, v_o/v_i, has a third order characteristic with three real poles, one real zero and a pair of high frequency complex conjugate zeros. All roots are located on the left-half of the s-plane. The DM DC gain, without simplifications, is more direct and is given by

$$A_{DC} = \frac{g_{m22}+g_{m23}}{g_{o22}+g_{o23}} \times \frac{g_{m12}(g_{m13}+g_{o13}+g_{o14}+g_{mb13})+g_{m13}g_{o14}}{g_{o12}(g_{m13}+g_{o13}+g_{o14}+g_{mb13})+g_{o13}g_{o14}}. \tag{4.11}$$

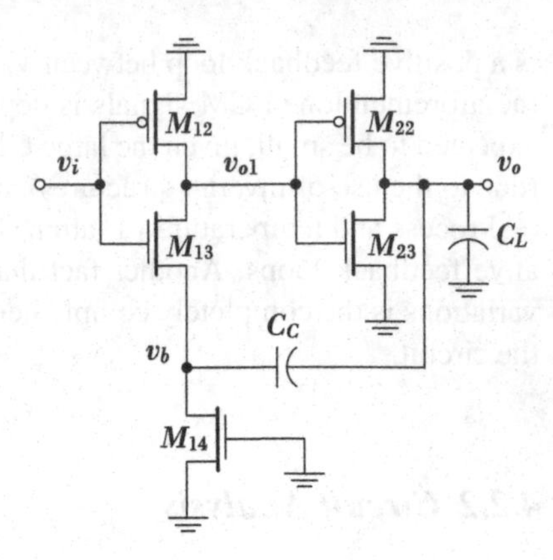

Fig. 4.14 Small-signal differential-mode half-circuit of the amplifier

From the nodal current equations it is possible to extract a behavioural signal path model which gives large insight in the small-signal behaviour of the amplifier [99]. This model is illustrated in Fig. 4.15 (the minor textual generalizations mentioned above are used). C_{o1}, C_o, and C_b represent the capacitance on nodes v_{o1}, v_o, and v_b, respectively and g_B represents the conductance of node v_b (refer to Fig. 4.14 for node names). Through Fig. 4.15 it is possible to verify:

- the DC gain path (by nulling all capacitors), represented by the direct path (solid line);
- the feedback loop created by the compensation capacitor, C_C. The Miller effect through parasitic capacitance C_{gd2}, represented by the dashed line;
- feedforward paths through C_{gs13} and C_C, represented by the dotted line;
- the poles (block with a low-pass filter function) and zeros (block with a high-pass filter function) of the TF. In other words, the order of the TF, as mentioned above, 3rd order.[4]

As mentioned before, the DM TF is rather complex to maintain a useful level of accuracy. Instead, it is more interesting to obtain insight in the role of each parameter (g_m, g_o, and capacitors) of the circuit. To provide this, pole-zero position diagrams are used [92]. These diagrams show the tendency of AC performance parameters (GBW, open-loop DC gain, etc.) when one circuit parameter is swept and the others are kept constant [92]. No simplifications need to be made on the TF. Due to the large number of circuit parameters, only some diagrams will be shown, while others will be discussed. Parameters g_{m13}, g_{m22} and C_C are chosen for the pole-zero position diagrams which are shown in Fig. 4.16. The parameter values used are $2g_{m12,13} = g_{m22,23} = 2.5\,\text{mS}$, $2g_{o12,13} = g_{o22,23} = 60\,\mu\text{S}$, $g_{o14} = 500\,\mu\text{S}$, $C_{db12,13,22,23} = 50\,\text{fF}$, $C_{gd12,13,22,23} = 25\,\text{fF}$, $C_{gs12,13,22,23} = 125\,\text{fF}$, $C_C = 500\,\text{fF}$, and $C_L = 4\,\text{pF}$. Each diagram illustrates the variation of the poles, low-frequency zero, unity-gain frequency (f_u) and the product of the open-loop DC gain, A_{DC}, and the dominant

[4] Generally, an N-node system (excluding DC nodes and input node) has N initial conditions, which correspond to N poles [71]. In this case, we have a 3-node system, hence a 3-pole system.

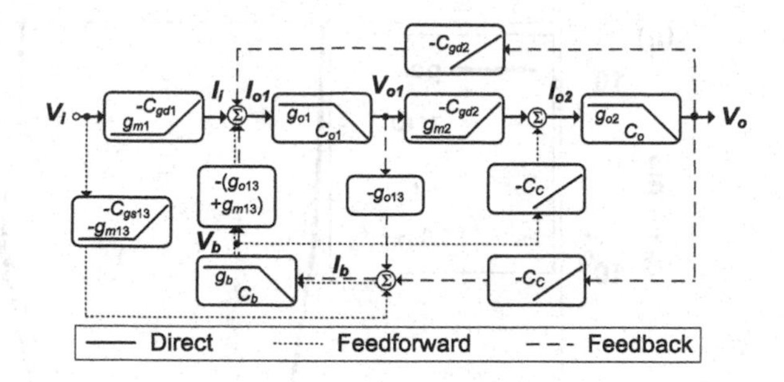

Fig. 4.15 Behavioural signal path model of the differential-mode half-circuit

pole, f_d, $A_{DC}f_d$.[5] A_{DC} is the horizontal distance between f_d and $A_{DC}f_d$ (see Fig. 4.16c).

Regarding g_{m13}, as it increases, A_{DC} increases slightly. For a specific value of g_{m13}, the nondominant pole can be cancelled by the low frequency zero. To avoid an oscillatory step response in closed-loop form (due to low phase margin, PM), a minimum value of g_{m13} should be kept. Regarding g_{m22} (and g_{m23} which has a similar effect on the TF), for rising values, the DC gain increases and a pole-splitting effect is noticeable. There is an optimum value for g_{m22} which maximizes the amplifier's settling time. Finally, C_C produces the expected pole-splitting effect and does not affect the DC gain. The optimum value is found just before the two real nondominant poles become complex (point at which the nondominant poles meet). At this point the PM is sufficient and the low frequency zero does not have a strong effect on the amplifier's step response. High values (>800 fF) cause complex poles (with a low damping factor) and the effect of the low frequency zero becomes pronounced [119]. Figure 4.17 depicts the variation of the phase margin and damping factor (ζ) of the nondominant poles in relation to C_C. The data points marked by a plus symbol are for PM > 60°. It can be seen that different values of the damping factor are achieved, i.e., the damping factor decreases with increasing C_C. Through MATLAB closed-loop step response simulations on each of the data points (i.e., for each C_C) it can be seen that the settling time increases as the damping factor decreases. Therefore, although PM> 60°, special attention must be given to ζ as it plays a fundamental role in the opamp's settling [119].

Regarding the remaining circuit parameters, all g_o have a similar effect on the TF: when increased they reduce A_{DC} and only affect the dominant pole. Transconductance g_{m12} has no effect on the poles and for rising values, A_{DC} increases. For values of $g_{m12} > 2$ mS the circuit becomes unstable with a very low PM. No parasitic capacitors have such pronounced effects worth mentioning. Some of the conclusions drawn for the above analysis are based on MATLAB function simulations (step, bode, damp, margin, etc.).

[5] $A_{DC}f_d$ = GBW only in single pole amplifiers or when a sufficient phase margin (PM > 60°) is obtained for multi-pole amplifiers.

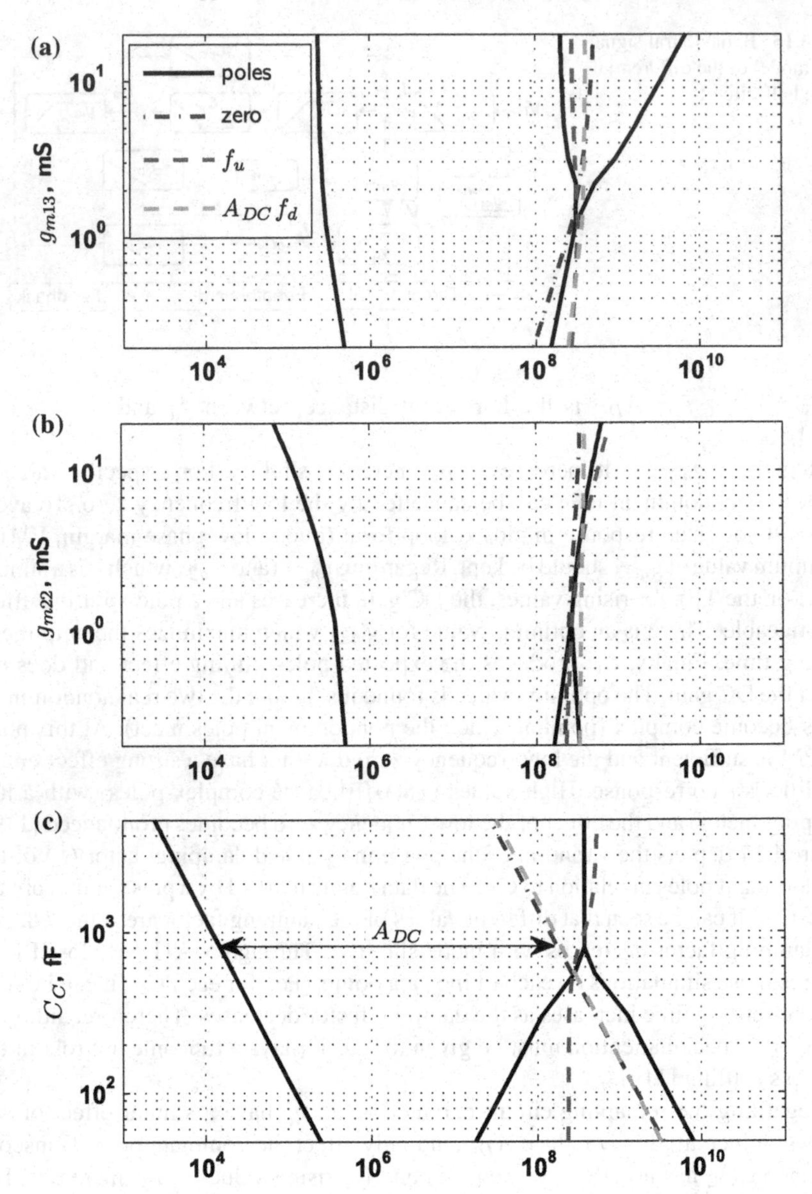

Fig. 4.16 DM pole-zero position diagrams for parameters: **a** g_{m13}, **b** g_{m22}, and **c** C_C

4.2.2.2 Common-Mode Feedback Analysis

The small-signal equivalent circuit for CMFB analysis is shown in Fig. 4.18 (v_{cm2} represents the input signal and v_{ocm} the output signal). It is evident that there are many paths from the input to the output: a path through M_{21} and M_{24} (two gain

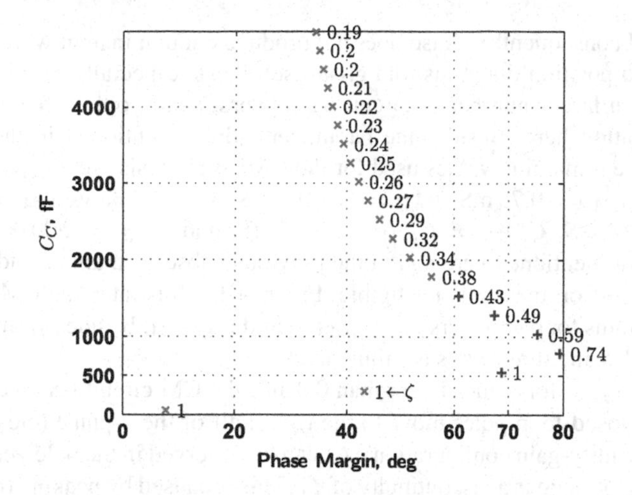

Fig. 4.17 Phase margin and nondominant pole damping factor (ζ) variation with C_C

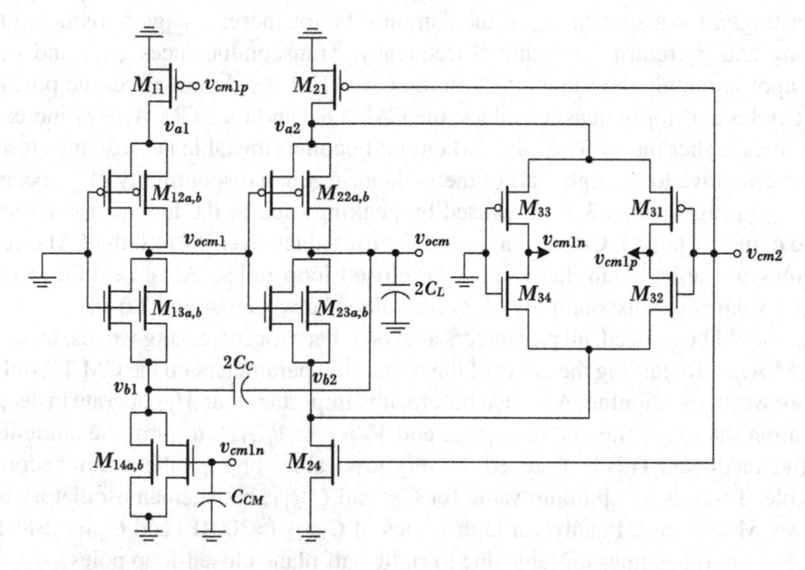

Fig. 4.18 Small-signal common-mode feedback equivalent circuit of the amplifier

stages from input to output), a path through v_{cm1p} (four gain stages) and one through v_{cm1n} (also four gain stages). At high frequencies, the v_{cm1n} path reduces to two gain stages because the CM signal is coupled through $2C_C$ to the output. Given the many poles in the CM path, it is expected that the CM circuit be slower than the DM circuit.

As can be seen, the CMFB equivalent circuit of the amplifier contains 8 signal nodes (excluding v_{cm2}) which correspond to 8 poles, and in this case, 8 zeros. It was not possible to extract the TF with a sufficient level of representation, even with simplifications. The behavioural signal path model becomes too complex and difficult to follow, and consequently, it also does not produce enough insight when analysed. The pole-zero position diagrams find great usefulness, especially in this situation. Given the even larger number of circuit parameters, again, only a few are chosen for representation here. Besides the parameter values mentioned in the previous subsection, the remaining values used for the CMFB analysis are $2g_{m14} = g_{m11} = 1.25\,\text{mS}$, $g_{m21,24} = 0.75\,\text{mS}$, $g_{m31-34} = 0.1\,\text{mS}$, $1.5g_{o11} = g_{o21,24} = 750\,\mu\text{S}$, $g_{o31-34} = 200\,\mu\text{S}$, $C_{gd11} = C_{gs11,14} = 350\,\text{fF}$, and $C_{CM} = 500\,\text{fF}$. All other parameters not mentioned here and in the previous subsection are considered zero, since their effect on the TF is negligible. Figure 4.19 illustrates the CM pole-zero position diagrams for parameters g_{m14}, $g_{m21,24}$ and g_{m31-34}. Besides f_u and $A_{DC} f_d$, only three poles and three zeros are illustrated.

Regarding g_{m14}, for values larger than 0.8 mS, the CM circuit becomes unstable because the closed-loop poles move to the right half of the s-plane (closed-loop is considered as unity-gain configuration). As can be observed in the pole-zero position diagram of g_{m14}, there is a discontinuity of f_u. This is caused by peaking (observable in the Bode diagrams) due to a low damping factor of the nondominant complex poles. Interestingly, for a specific g_{m14}, the damping factor increases again, reducing the peaking and f_u returns to its initial frequency. Transconductances g_{m21} and g_{m24} have approximately the same effect on the CM TF. They do not affect the position of the poles and for increasing values, the CM A_{DC} and thus CM $A_{DC} f_d$ increase. For values higher than 2.3 mS, the CM circuit becomes unstable because the closed-loop poles move to the right half of the s-plane. Again a discontinuity of f_u occurs, for $g_{m21,24}$ higher than 3.7 mS, caused by peaking (due to the low damping factor of the complex poles). Concerning g_{m31-34}, for values over 0.2 mS, the CM circuit becomes unstable, again due to positive closed-loop poles. Also, peaking occurs which explains the discontinuity of f_u for values between 0.4 and 1.6 mS.

As should be noticed, all parameters analysed here for increasing values, increase the CM A_{DC}. Regarding the effect of the remaining parameters on the CM TF, only a few are worth mentioning. As stated before, it is important that M_{11} operate in deeper saturation than M_{14} (i.e., $g_{m11} > g_{m14}$ and $V_{od11} < V_{od14}$), to limit the amount of positive feedback. This is observed for very low values of g_{o14}, the circuit becomes unstable. There is an optimum value for C_C and C_{gs11}, between an oscillatory and a slow CM response. Finally, for high values of C_{gd11} (>300 fF) and C_{gd14} (>80 fF) the CM circuit becomes unstable due to right-half plane closed-loop poles.

As mentioned before, the CM circuit is more likely to be slower than the DM circuit. This is well observed by comparing the average value of $A_{DC} f_d$ of the pole-zero diagrams of Fig. 4.16 (except for C_C) with the diagrams of Fig. 4.19. The average value of the DM $A_{DC} f_d$ is situated between 200–300 MHz, while the CM $A_{DC} f_d$ is below 100 MHz. Again, some of the conclusions drawn here are based on MATLAB simulations (step, bode, damp, margin, etc.).

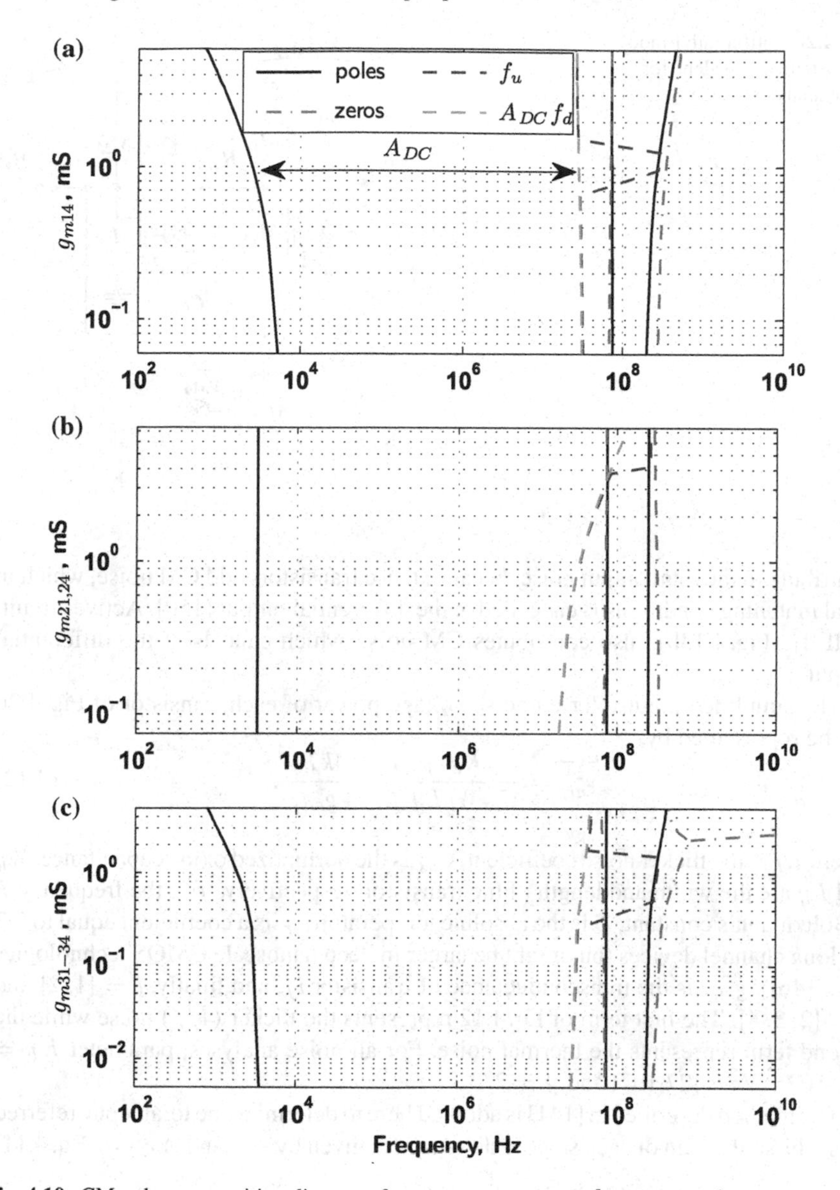

Fig. 4.19 CM pole-zero position diagrams for parameters: **a** g_{m14}, **b** $g_{m21,24}$, and **c** g_{m31-34}

4.2.2.3 Noise

The noise of amplifiers can be a limiting factor in many applications, thus the need to characterize the noise performance of this amplifier. The circuit of Fig. 4.20 is used to analyse the noise of the amplifier [141], where v_{nie} represents the input equivalent noise voltage and v_{no}, the output noise voltage. Transistors M_{11}, M_{21} and M_{24} do not

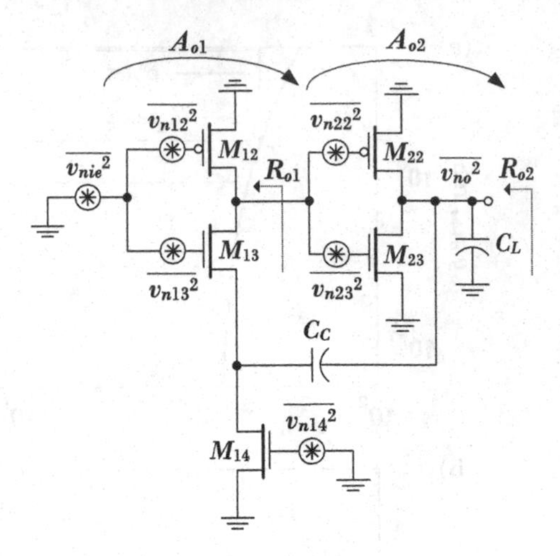

Fig. 4.20 Differential-mode half-circuit equivalent for noise analysis

contribute to the total output noise, because these transistors add CM noise, which in ideal matching conditions is cancelled at the differential output [150]. Active circuit, $CMFB_1$ (Fig. 4.13b), also contributes CM noise which cancels at the differential output.

The simplified input-referred noise voltage power of each transistor of Fig. 4.20 can be represented by,

$$\overline{v_{nij}^2} = \frac{K_F}{C_{ox} W_{ij} L_{ij}} \frac{1}{f} + \frac{4kT\gamma}{g_{mij}}. \tag{4.12}$$

where K_F is the flicker noise coefficient, C_{ox} is the normalized oxide capacitance, W_{ij} and L_{ij} are the width and length of the transistor, respectively, f is the frequency, k is Boltzmann's constant, T is the absolute temperature, γ is a coefficient equal to 2/3 for long channel devices (but might be larger in deep nanoscale CMOS technologies [141,146]), g_{mij} is the transconductance of transistor ij, and finally, $i = \{1, 2\}$ and $j = \{2, 3, 4\}$. The first term of Eq. 4.12 represents the flicker $(1/f)$ noise while the second term represents the thermal noise. For all noise analyses, parameter $K_F = 4 \times 10^{-23}\ V^2F$ [27,107] and $\gamma = 2/3$.

The method described in [141] is adopted here to determine the total input-referred noise. First, the gain of each stage is determined, given by A_{o1} and A_{o2} (see Eq. 4.11)

$$A_{o1} = -\frac{g_{m12}(g_{m13} + g_{o13} + g_{o14}) + g_{m13}g_{o14}}{g_{o12}(g_{m13} + g_{o13} + g_{o14}) + g_{o13}g_{o14}}, \tag{4.13a}$$

$$A_{o2} = -\frac{g_{m22} + g_{m23}}{g_{o22} + g_{o23}}, \tag{4.13b}$$

where $g_{mb13} = 0\,\mathrm{S}$, for a more simplified analysis. Then, the impedance seen from the output of each stage is determined,

$$R_{o1} = \frac{g_{m13} + g_{o13} + g_{o14}}{g_{o12}(g_{m13} + g_{o13} + g_{o14}) + g_{o14}g_{o13}}, \tag{4.14a}$$

$$R_{o2} = \frac{1}{g_{o22} + g_{o23}}. \tag{4.14b}$$

Each noise voltage source is referred to the amplifier input by converting to its respective noise current source, multiplying the latter by the squared output impedance of that stage and dividing it by the appropriate squared gain (A_{o1} for transistors of stage 1 or $A_{o1}A_{o2}$ for transistors of stage 2). In other words, for M_{12}, M_{13} and M_{14}, their input referred noise is $v_{n1j}^2 g_{m1je}^2 R_{o1}^2 / A_{o1}^2$ and for M_{22} and M_{23} it is $v_{n2j}^2 g_{m2j}^2 R_{o2}^2 / (A_{o1}^2 A_{o2}^2)$, where v_{nij} is given by Eq. 4.12 and g_{m1je} is described in [141] as being an equivalent transconductance of transistor M_{1j} ($j = \{2, 3, 4\}$). This type of transconductance needs to be defined for amplifier stages with source degeneration or when the output impedance is not measured at the transistor's terminals, as is the case for M_{13} and M_{14}, respectively. The equivalent transconductances of g_{m13} and g_{m14} are given by

$$g_{m13e} = \frac{g_{m13}g_{o14}}{g_{m13} + g_{o13} + g_{o14}}, \tag{4.15a}$$

$$g_{m14e} = \frac{g_{m14}(g_{m13} + g_{o13})}{g_{m13} + g_{o13} + g_{o14}}, \tag{4.15b}$$

respectively. The total input-referred noise spectral density is the sum of each input referred noise voltage (two times to account for the differential circuit) and is given by

$$\overline{v_{nie}^2} = 2 \times \left[\frac{R_{o1}^2}{A_{o1}^2} (\overline{v_{n12}^2} g_{m12}^2 + \overline{v_{n13}^2} g_{m13e}^2 + \overline{v_{n14}^2} g_{m14e}^2) + \frac{R_{o2}^2}{A_{o1}^2 A_{o2}^2} (\overline{v_{n22}^2} g_{m22}^2 + \overline{v_{n23}^2} g_{m23}^2) \right]. \tag{4.16}$$

By simplifying Eqs. 4.15a, 4.15b, and 4.16 ($g_m \gg g_o$), it is possible to extract the devices that contribute most noise to the amplifier. These devices are M_{12} and M_{14}, because $g_{m14e} \approx g_{m14}$, $g_{m13e} \approx g_{o14}$, and the noise of $M_{22,23}$ is divided by a larger gain (when referred to the input). Considering only thermal noise and the mentioned simplifications ($g_m \gg g_o$), Eq. 4.16 can be written in a more common way

$$\overline{v_{ni_{Thermal}}^2} = \frac{8kT\gamma}{g_{m12}} \left(1 + \frac{g_{o14}^2}{g_{m12}g_{m13}} + \frac{g_{m14}}{g_{m12}} \right) \approx \frac{8kT\gamma}{g_{m12}} \left(1 + \frac{g_{m14}}{g_{m12}} \right), \tag{4.17}$$

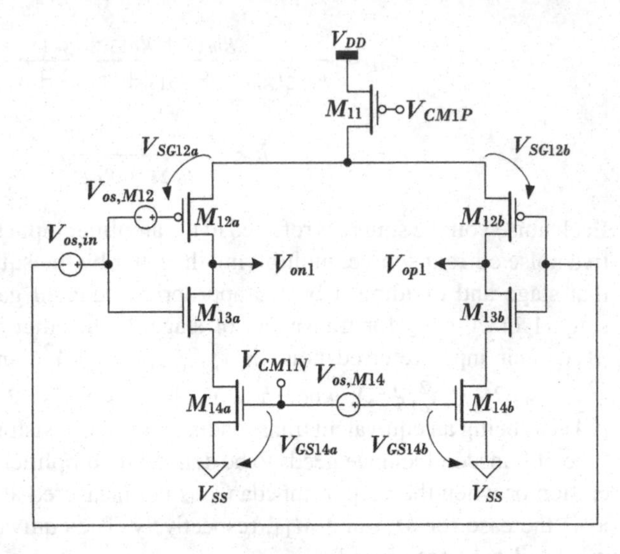

Fig. 4.21 Differential-mode equivalent for input-referred offset analysis

where $1 + g_{m14}/g_{m12}$ yields the thermal noise excess factor. To determine the integrated output noise, the thermal noise term of Eq. 4.17 is multiplied by the amplifier's bandwidth, $\mathrm{GBW}\pi/2 \approx g_{m12}/(2\pi C_C)\pi/2$ (considering a first-order roll off)

$$v_{noThermal} = \sqrt{\frac{8kT\gamma}{g_{m12}}\left(1 + \frac{g_{m14}}{g_{m12}}\right)\mathrm{GBW}\frac{\pi}{2}} \approx \sqrt{2\left(1 + \frac{g_{m14}}{g_{m12}}\right)\gamma\frac{kT}{C_C}}. \quad (4.18)$$

4.2.2.4 Input Offset

Similar to what occurs with noise, amplifiers have no way of eliminating offset. Naturally it is possible to reduce and minimize offset (like noise) to the extent where it is negligible and does not limit performance. Therefore, it is important to analyse this amplifier to understand which devices contribute to the offset.

All analyses carried out for this amplifier (except this one) assume the amplifier is perfectly symmetric, i.e., identical transistors (e.g., M_{12a} and M_{12b}) are perfectly matched. The manufacturing process introduces a number of uncertainties, which lead to small variations across the chip, ultimately resulting in mismatch between otherwise identical devices. As explained in the previous section, mismatch may exist in various transistor parameters, namely, V_{TH}, W, L, C_{ox}, and μ. Mathematically, these mismatches are given by Eqs. 4.7a, 4.7b, which result in Eq. 4.8.

Figure 4.21 depicts the circuit for input-referred offset analysis. Only the input stage is considered given that the offset of the output stage will be much smaller because it is divided by a higher gain, when referred to the input. As mentioned before, mismatch occurs between two devices that are, otherwise, identical. Therefore, M_{11} can be immediately discarded from this analysis. Similar to the noise analysis, M_{13}

plays a minor role in the offset due to its source degeneration. Therefore, the following expressions are for the $V_{os,M12}$ and $V_{os,M14}$ offset voltages, which have already being simplified using Eq. 4.8

$$V_{os,M12} = V_{SG12b} - V_{SG12a} = \frac{(V_{SG} - V_{TH})_{12}}{2}\left(\frac{\Delta(W/L)}{W/L}\right)_{12} + \Delta V_{TH,12}, \tag{4.19a}$$

$$V_{os,M14} = V_{GS14b} - V_{GS14a} = \frac{(V_{SG} - V_{TH})_{14}}{2}\left(\frac{\Delta(W/L)}{W/L}\right)_{14} + \Delta V_{TH,14}. \tag{4.19b}$$

Referring the above described offset voltages to the amplifier's input, $V_{os,in}$, and only considering the ΔV_{TH} term, given that it is the largest offset contributor, the total input-referred offset is given by

$$V_{os,in} \approx \sqrt{\left(\frac{A_{V_{TH,P}}}{\sqrt{(WL)_{12}}}\right)^2 + \left(\frac{A_{V_{TH,N}}}{\sqrt{(WL)_{14}}}\frac{g_{m14}}{g_{m12}}\right)^2}. \tag{4.20}$$

The method used in the noise analysis for referring the various noise voltage sources to the amplifier's input, can also be applied here. In this case, instead of noise voltage sources, offset voltage sources are employed.

4.2.2.5 Slew Rate

One important design aspect that should always be taken into consideration is the slewing of the output. Slewing occurs in the presence of large input signals, which, due to current limitations of the amplifier, causes the output to ramp at a constant rate, the slew rate (SR), independently of the input signal, thus causing nonlinear distortion [92,141]. Besides distortion, slewing has a direct influence on the settling time [7]. Figure 4.22 shows the equivalent circuits used to analyse the positive SR, SR^+, and negative SR, SR^-, of the amplifier. The following describes the situation in one half of the fully differential circuit; naturally the opposite occurs in the other half of the circuit. Figure 4.22a shows the case of a large positive input signal. In this case M_{12a} turns off (shown by the dashed lines), limiting the current to M_{13a} and M_{14a}. Node V_{on1} is thus pulled down, leaving M_{13a} in the deep triode region. As node V_{on1} decreases, M_{23a} is turned off, pushing current I_{B2} into M_{22a}. Current I_{B2} must be large enough to charge the load capacitor and still source current I_{D14a} through C_{Ca}. Given that node V_{ba} is not a fixed voltage in the sense that it varies during slewing, the positive SR is determined by the load capacitor, C_L, and the output current, I_{CL}. C_L is charged at a rate given by

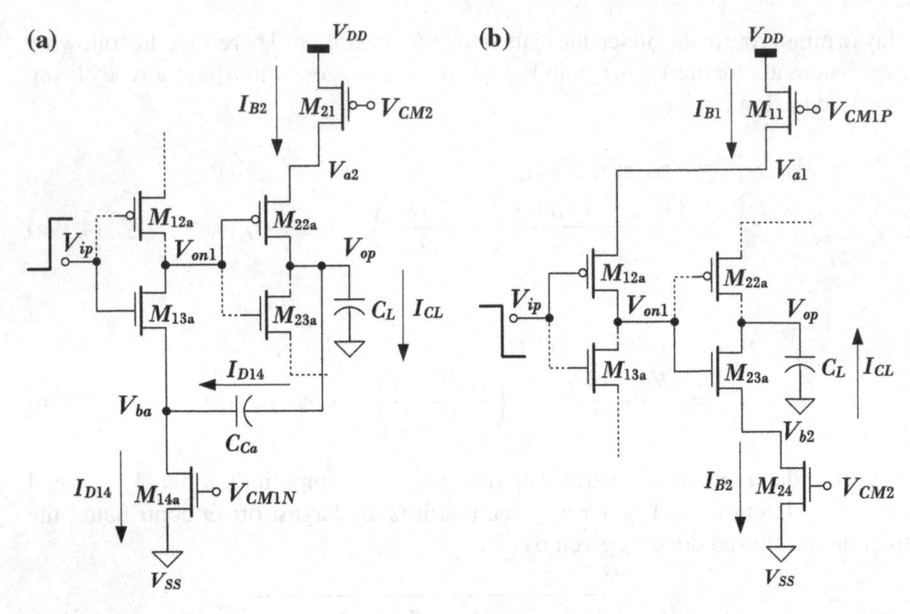

Fig. 4.22 Differential-mode half-circuit equivalents for slew rate analysis: **a** positive input (SR$^+$). **b** Negative input (SR$^-$)

$$\mathrm{SR}^+ = \frac{I_{CL}}{C_L} = \frac{I_{B2} - I_{D14a}}{C_L} = \frac{I_{B2} - \frac{I_{B1}}{2}}{C_L}. \tag{4.21}$$

During negative slewing, which occurs for a large negative input as shown in Fig. 4.22b, M_{13a} is turned off pulling node V_{on1} high. As node V_{on1} rises, M_{22a} turns off, cutting the current to M_{23a}. Current source M_{24} still tries to sink current I_{B2}, which can only come from C_L, therefore, discharging it at a rate given by

$$\mathrm{SR}^- = \frac{I_{CL}}{C_L} = \frac{I_{B2}}{C_L}. \tag{4.22}$$

To simplify the carried out analysis, it is considered that when a transistor turns off, no current flows through it. Moreover, the analysis did not consider the circuit of CMFB$_1$ which is connected to the main amplifier through nodes V_{a2} and V_{b2}. Depending on the biasing conditions, sizing and large signal input, the circuit of CMFB$_1$ can source current to the amplifier's output stage, contributing to the total SR. This situation is depicted in Fig. 4.23. For example, during a large input signal transient, V_{a2} can decrease to below V_{CMIP}, where M_{31} would have its source and drain terminals exchanged, therefore, instead of sourcing current to M_{32} (now turned off), it would source current to the output stage. The current flows from V_{DD} into node V_{CM1P} through source-gate capacitance, C_{sg11}, of M_{11}.

Finally, these analyses consider that M_{21} and M_{24} operate in the saturation region. It is more likely that one of these transistors operate in the triode region (depending

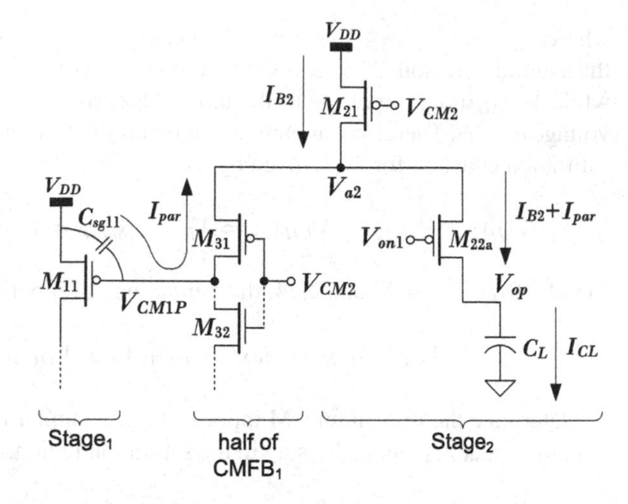

Fig. 4.23 CMFB$_1$'s participation in increasing the total SR

on V_{CM2}) and in this case the output current is defined by M_{22} (instead of M_{21}) for SR$^+$ and M_{23} (instead of M_{24}) for SR$^-$.

In [106] a comparison is made between using self-biasing and constant biasing, applied to amplifiers, and its relation to SR performance. The authors of [106] concluded that the self-biased SR was lower by about 10 %. Simulations of the proposed amplifier (including CMFB$_1$) show that the SR is approximately 11 % lower when using self-biasing, which is in close agreement with [106]. Moreover, with constant current sources, the amplifier's SR$^+$ and SR$^-$ are equal, but with self-biasing, SR$^+$ is slightly higher than SR$^-$.

4.2.2.6 Input Range

To determine the input CM range (CMR) of the amplifier, it is necessary to determine the maximum and minimum CM voltages that maintain the devices of the input stage in saturation [92,141]. The circuit of Fig. 4.24 will be used for this analysis. Maintaining their operation in the saturation region will guarantee their desired transconductance and therefore the performance of the amplifier (A_{DC}, GBW, etc.). For the following analyses, V_{THij} and V_{ODij} denote transistor ij's threshold and overdrive voltage ($V_{GS} - V_{TH}$), respectively. To simplify the analyses, the threshold voltage of NMOS and PMOS transistors are considered equal (in absolute value) and denoted by V_{TH}. For the NMOS devices of the input stage, the input CM voltage, $V_{in,CM}$, must guarantee, at least, the gate-source voltage of M_{13}, V_{GS13}, and M_{14}'s overdrive voltage, V_{OD14}, which can be written as,

$$V_{in,CM,MIN} = V_{GS13} + V_{OD14} = 2V_{OD} + V_{TH}, \tag{4.23}$$

where $V_{GS13} = V_{OD13} + V_{TH13}$. Simultaneously, as $V_{in,CM}$ is reduced, M_{12} may leave the saturation region. M_{12} leaves saturation because as $V_{in,CM}$ falls, V_{a1} falls as well while V_{OCMI} (output voltage of the input stage) rises, reducing V_{SD12}, source-drain voltage of M_{12}. Therefore, another equation may determine $V_{in,CM,MIN}$, which is the saturation equation for M_{12}, given by

$$V_{SD12} \geq V_{SG12} - |V_{TH12}| \Rightarrow V_{in,CM,MIN} = V_{OCM1} - V_{TH}. \tag{4.24}$$

Finally, from Eqs. 4.23 and 4.24, the minimum CM input voltage is given by,

$$V_{in,CM,MIN} = \max\{2V_{OD} + V_{TH}, V_{OCM1} - V_{TH}\}. \tag{4.25}$$

To determine the maximum CM input voltage, a similar reasoning can be used. To guarantee that M_{11} and M_{12} stay in the saturation region,

$$V_{in,CM,MAX} = V_{DD} - V_{OD11} - V_{SG12} = V_{DD} - 2V_{OD} - V_{TH}. \tag{4.26}$$

As $V_{in,CM}$ rises, V_{b1} also rises while V_{OCM1} falls, driving M_{13} into the triode region. To guarantee that M_{13} stays in the saturation region, we use its saturation equation and arrive at

$$V_{in,CM,MAX} = V_{o1} + V_{TH13} = V_{OCM1} + V_{TH}. \tag{4.27}$$

From Eqs. 4.26 and 4.27, the maximum CM input voltage is given by

$$V_{in,CM,MAX} = \min\{V_{DD} - 2V_{OD} - V_{TH}, V_{OCM1} + V_{TH}\}. \tag{4.28}$$

The CMR that maintains the transistors of the input stage in saturation is comprised between Eqs. 4.25 and 4.28, given by

$$\begin{aligned} \max\{2V_{OD} + V_{TH}, V_{OCM1} - V_{TH}\} &\leq V_{in,CM} \\ &\leq \min\{V_{DD} - 2V_{OD} - V_{TH}, V_{OCM1} + V_{TH}\}. \end{aligned} \tag{4.29}$$

This means that the input stage cannot process a rail-to-rail input as could have been expected. This is because the current through the input stage depends on the correct operation of all transistors. For example, as the CM input rises, the NMOS devices should conduct more, but the PMOS devices progressively cut off, cutting the current to the NMOS devices, thus reducing their transconductance.

To explain the main difference from this kind of input stage to the ones presented in [18, 44, 45, 74] (complementary input stages), which have a rail-to-rail input range, the same example is used: rising CM input. Complementary input stages have separate NMOS and PMOS inputs, therefore, as the CM input rises the NMOS conducts "freely" as its current does not depend on the operation of the PMOS input. The PMOS input will eventually cut off, but will not affect the NMOS input. A similar analogy can be made for falling CM inputs.

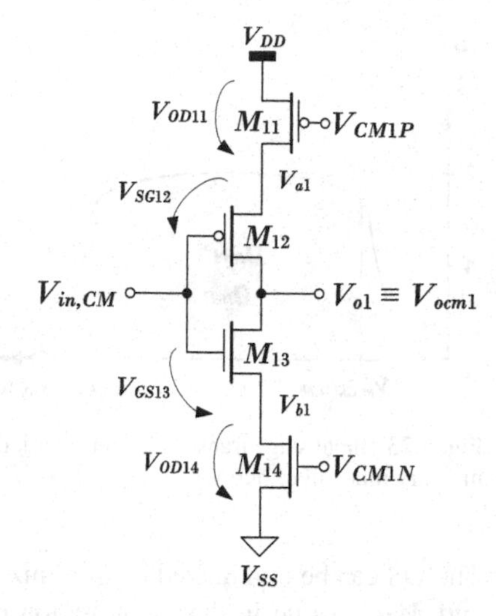

Fig. 4.24 Input stage for input CM range analysis

A drawback of the complementary input stage of the proposed amplifier is the G_m variation over the CMR, as shown in Fig. 4.25. Here, G_m denotes the sum of the PMOS and NMOS transistors' transconductances, g_{mP} and g_{mN}, respectively (no source degeneration of M_{13} is considered for this analysis). Figure 4.25b shows that there is always one input, thus one g_m, conducting, which does not happen for the proposed amplifier. The proposed amplifier's G_m variation is larger than that of the circuits presented in [18, 44, 45, 74]. It is imperative to mention that the proposed amplifier was not designed to have a constant-g_m rail-to-rail input.

Concerning the self-biasing voltages V_{CM2}, V_{CM1P}, and V_{CM1N}, these are approximately constant over the entire CMR of the amplifier. They maintain approximately constant due to the fact that both stages of the amplifier and $CMFB_1$ are inverter-based, which have larger CM input ranges than single transistor gain stages.

The minimum supply voltage for the input stage to operate correctly is given by

$$V_{DD,MIN} = 2V_{GS} + 2V_{OD} = 2V_{TH} + 4V_{OD}. \tag{4.30}$$

4.2.2.7 Output Swing

To analyse the amplifier's output swing (OS), Fig. 4.26, all transistors of the output stage are considered operating in the saturation region. In this case the differential OS is given by

$$OS_{sat} = 2(V_{DD} - V_{OD21} - V_{OD22} - V_{OD23} - V_{OD24}) = 2(V_{DD} - 4V_{OD}). \tag{4.31}$$

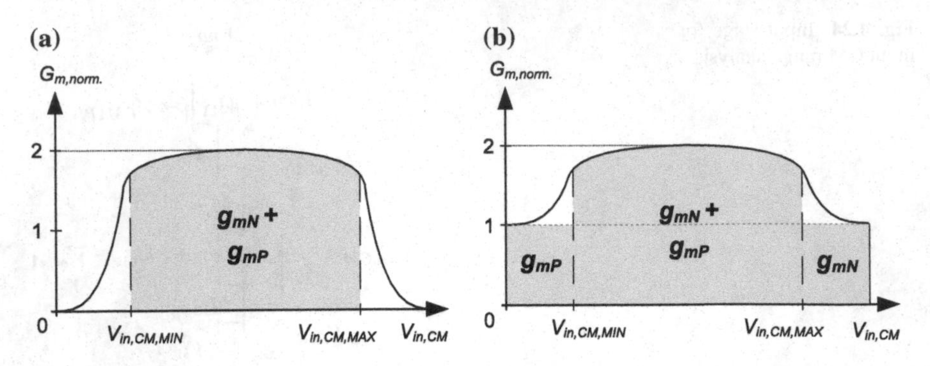

Fig. 4.25 Input stage transconductance variation over the CMR: **a** proposed amplifier. **b** Complementary input amplifiers

The OS can be maximized by minimizing the overdrive voltages of M_{21-24}. If M_{21} and M_{24} operate in the triode region (most likely situation to occur, as shown in [10]), the OS can be rewritten as

$$\mathrm{OS}_{sat} = 2(V_{DD} - V_{OD22} - V_{OD23} - I_{B2}(R_{P21} + R_{N24})) = 2(V_{DD} - 2V_{OD} - 2V_{DS}). \quad (4.32)$$

where I_{B2} is the output stage's current, and R_{P21} and R_{N24} represent the resistance of M_{21} and M_{24} (transistors in the triode region), respectively. In this case a higher OS can be achieved given that the resistors can be minimized, thus shifting V_{a2} and V_{b2} toward the supply rails. For the OS calculations presented above, no saturation margin (to account for process and parameter variations) was subtracted. When designing the amplifier, a margin of at least 40 mV[6] per transistor (in the saturation region) should be used.

The minimum supply voltage for the output stage to operate correctly is given by

$$\begin{cases} V_{DD,MIN_{sat}} = \frac{\mathrm{OS}}{2} + 4V_{OD} \\ V_{DD,MIN_{triode}} = \frac{\mathrm{OS}}{2} + 2V_{OD} + 2V_{DS}. \end{cases} \quad (4.33)$$

The minimum supply voltage considering both stages of the amplifier is given by

$$V_{DD,MIN} = \max\{2V_{TH} + 4V_{OD}, \frac{\mathrm{OS}}{2} + 2V_{OD} + 2V_{DS}\}. \quad (4.34)$$

4.2.2.8 Considerations on the Amplifier's Class

One important question arises concerning the class of the proposed amplifier given its self-biasing and inverter-based architecture. The analog inverter is the most used example when characterizing class-AB amplifiers. Usually those examples employ

[6] This value was extracted from PVT corner simulations of the amplifier that is presented in Sect. 6.1.

Fig. 4.26 Output stage for output swing analysis

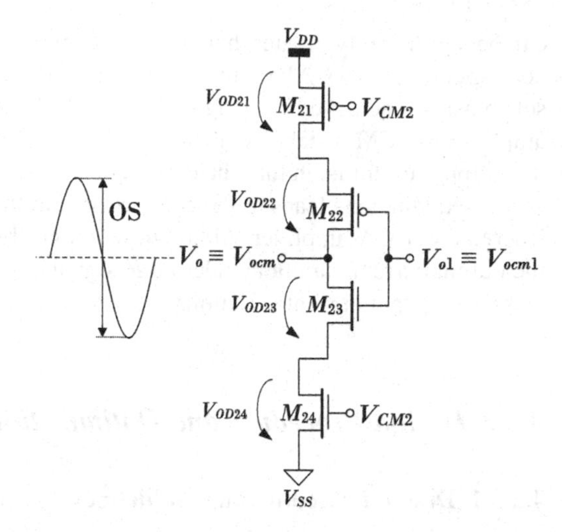

a simple inverter with no other transistors [146]. In the inverter case (class-AB), the quiescent biasing current, I_Q, is usually made small (for power efficiency reasons). The peak output current, I_{peak}, can be much larger than I_Q because the transistors' V_{GS} (which depends on the input voltage) can be rail-to-rail, so simple inverter amplifiers can source/sink a much larger current than the quiescent current. References [18, 55, 124, 163] are examples of class-AB amplifiers where the peak current is much larger than the quiescent current ($I_{peak}/I_Q > 10$). Actually, this is one of the main requisites in class-AB amplifiers [146].

In the proposed amplifier, both stages use inverters as their input devices, but they are also accompanied by current sources which define the current in each stage. For the remainder of this discussion, only the output stage will be of relevance. Transistors M_{21} and M_{24} define the current of the output stage and are biased by the output common-mode (CM) voltage (V_{CM2}) of the amplifier, which, may not be constant (depending on CM variations). Although $M_{22a,b}$ and $M_{23a,b}$ operate exactly like an inverter, in the sense that, depending on their input, current flows more through one transistor than the other, the total available current is always limited to the current sourced/sunk by M_{21} and M_{24}, in other words, $I_{peak}/I_Q \approx 1$. Nevertheless, under large signal conditions V_{CM2} may vary, leading to a variation of the output current which may be made larger than I_Q, precluding this amplifier as purely class-A. Class-B can be ruled out since the quiescent biasing current is not null. Large input signal simulations of the proposed amplifier show that the total output current varies less than 15 % around the quiescent current (this is mainly due to CM variations). So, in fact there is some variation of the output current for large input signals, but $I_{peak}/I_Q < 1.15 \ll 10$.

Regarding the self-biasing part of the issue, in [10] and again exampled in [55], self-biasing is accomplished by using a differential-mode (DM) signal (usually one of the outputs). In this case the switching currents (under large signal conditions) can be significantly greater than the quiescent biasing current [10,55], which would correspond to a class-AB output. In [180] and as is the case of the proposed amplifier, self-biasing is achieved by using a CM signal, which suffers less variations (unless the amplifier has CM instability) than DM signals. With a more steady biasing voltage, the output current is, in turn, held more constant.

All explanations leads to the conclusion that the proposed amplifier can be considered a class-A amplifier. Although it can not be considered purely of this class, the output current variation under large signal conditions is very small compared to class-AB output current variations.

4.2.3 Design Procedure and Optimization

4.2.3.1 Design Procedure and Guidelines

It is not an easy task, determining guidelines for a good and successful sizing of this amplifier, due to the complex TFs obtained for the DM and CM circuits. Although there is not a clear design procedure, some precautions and considerations can be extracted from the analyses previously carried out which may provide a good starting point. It is important to understand that this process is manual and iterative, and obtaining a good solution immediately is challenging.

The proposed design guidelines for this amplifier are as follows:

- The minimum value of C_C is mainly imposed by the kT/C thermal noise constraints which is set by the application where the amplifier is being used. C_C should not be too large to avoid complex nondominant poles and the effect of the low frequency zero. If complex poles are unavoidable, special attention must be given to their damping factor. If a good phase margin, $60° < \mathrm{PM} < 70°$, and complex poles are avoided, then a good time response can be achieved.
- M_{12} should be designed for high g_m and low g_o for high DC gain, i.e., large L and low V_{OD}. M_{12}'s transconductance should not be too large to avoid instability (this g_m does not affect the circuit's DM poles). M_{12} is the only transistor that does not affect the DM bandwidth. For low noise and low offset, this transistor should have a high transconductance. To increase the CM input range, V_{OD} should be low.
- Transistor M_{13} should have a large L or low V_{OD} for high DC gain. High g_{m13} aids the compensation, as it has a pole-splitting effect until a certain value, then the circuit becomes unstable. This transistor has a minor contribution in the amplifier's noise and offset. To increase the CM input range, this transistor should have a low V_{OD}.
- Transistors M_{22} and M_{23} should have low V_{OD} to increase the OS. They should have large L for high DC gain, but small L for high speed. These transistors can be used for additional compensation given their pole-splitting effect. Special care must be taken with the damping factor of possible complex nondominant poles

and the low frequency zero. The latter may have an undesirable effect on the step response of the amplifier.

- Transistor M_{11} should be biased in the saturation region with a V_{DS} that keeps M_{12} saturated. M_{11} should be sized to guarantee the current necessary for the g_m of transistors M_{12} and M_{13}. To increase the CM input range, this transistor should have a low V_{OD}.
- M_{14} should be designed always keeping in mind that, $g_{m14}/g_{o14} < g_{m11}/g_{o11}$, to limit the amount of positive feedback, to maintain CM stability. M_{14} should operate in the saturation region to maintain $V_{ba,b}$ a low impedance node. However, to maintain CM stability it should operate in a weak-to-moderate inversion region. Special care should be taken when sizing M_{14}, as this device is one of the largest noise and offset contributors, and is the only transistor that does not affect the DM GBW.
- Transistors M_{21} and M_{24} should be biased in the triode/saturation boundary region, with low V_{OD} to guarantee highest possible OS. These transistors should be sized to guarantee the current for M_{22} and M_{23}. The intrinsic gain of M_{21} and M_{24} ($g_{m21,24}/g_{o21,24}$) should be kept as equal and as low as possible, to avoid complex poles. For high SR, their channel widths should be made wider to sink/source larger currents.
- Transistors M_{31} to M_{34} should be biased to have $g_{m31-34}/g_{o31-34} \approx 1$ or less. Special attention should be devoted to M_{31} and M_{34} over PVT corners as their sizing may determine CM stability. The safest choice would be to bias them in a deep triode region.
- CM stabilization capacitor, C_{CM}, should not be too large to avoid a slow CM response and not too small to avoid an oscillatory response. A good trade-off is $C_{CM} = C_C$.
- Optimum channel length of 1.3-to-1.5$\times L_{min}$ should be used to maintain good insensitivity to PVT variations, minimizing short channel-length effects and, at the same time, maximizing speed.

A good starting point, assuming C_C is already set by thermal noise constraints, would be to set the Ls of the inverter pairs, $M_{12,13}$ and $M_{22,23}$, to $1.5 \times L_{min}$. Set the Ls of $M_{11,14}$ and $M_{21,24}$ to $2 \times L_{min}$. Set the Ls of the transistors of CMFB$_1$ to $1.5 \times L_{min}$. Set the Ws of all the PMOS devices K_N/K_P[7] times higher than the Ws of the NMOS devices. Finally, by following the enumerated design guidelines, the amplifier may be designed.

4.2.3.2 Design Methodology for Optimization

As an alternative to the manual design procedure, a software-based optimization platform, based on genetic algorithms (GA) is used. It will not be the objective of this section to describe the functionality of this tool, but rather to give a brief overview

[7] $K_{N,P}$ represents $\mu_{N,P} C_{ox}$.

of the optimizations carried out. More information on this tool can be found in [161]. Two different combinations were employed on the platform. Initially, the GA was combined with a frequency-domain analysis with results presented in [47]. Given that a prototype amplifier is to be integrated, a time-domain approach had to be followed.

Time-domain optimization can significantly simplify the calculus for circuit optimization of superior order topologies (more than three poles) and provide accurate results when compared to electrical simulation. With this methodology, besides current consumption, output swing and eventually noise, the only specification to consider is the settling time and the respective settling accuracy. Analyzing the step response of the circuit-under-optimization, it becomes relatively straightforward to obtain a single key performance indicator that encloses all the traditional frequency domain indicators, such as DC gain, GBW, β, and PM.

For the amplifier presented here, the GA individual format consists of the dimensions (W and L) of the transistors of the amplifier, as well as, the compensation capacitances. During optimization, some of the design constraints (as enumerated before) are taken into account to limit the search design space. Moreover, each optimization individual is classified by a fitness value, which results from the comparison between the behaviour of each circuit parameter and the desired specifications. Hence, each circuit parameter is compared by means of a partial fitness, f_i, which may assume three forms, depending on the desired type of optimization goal: maximize, minimize, or target (bounded) value, respectively represented by

$$f_i(p_i) = 1 - e^{-w_i \frac{p_i}{p_{i_{desired}}}}, \tag{4.35a}$$

$$f_i(p_i) = 1 - e^{-w_i \frac{p_{i_{desired}}}{p_i}}, \tag{4.35b}$$

$$f_i(p_i) = \frac{2}{e^{-w_i \frac{p_{i_{desired}} - p_i}{p_{i_{desired}}}} + e^{w_i \frac{p_{i_{desired}} - p_i}{p_{i_{desired}}}}}, \tag{4.35c}$$

where $p_{i_{desired}}$ represents the desired value, p_i is the current performance parameter value achieved by the individual being evaluated, and w_i represents the weight (or importance) assigned to the respective indicator in the fitness calculation.

Considering the frequency-domain approach, the following fitness function is employed

$$\text{fitness}_{\text{freq.}} = f_{A_{DC}} f_{\text{GBW}} f_{\text{PM}} f_{I_{total}} f_{\text{OS}}, \tag{4.36}$$

where f_i represents the partial fitness of each performance parameter defined below:

- $f_{A_{DC}}$, the DC gain, to be maximized,
- f_{GBW}, the gain-bandwidth product, to be maximized,
- f_{PM}, the phase margin, to be bounded,
- $f_{I_{total}}$, the total current consumption, to be minimized,
- f_{OS}, the output swing, to be maximized.

For the time-domain optimization, the fitness is simply given by

$$\text{fitness}_{\text{time}} = f_{Settling} f_{I_{total}} f_{\text{OS}}, \tag{4.37}$$

where $f_{Settling}$ is defined as the differential output voltage settling time for a specific settling accuracy, given a small-signal input step ($\approx$100 mV). Parameter $f_{Settling}$ is to be minimized. It should be clear that using the time optimization methodology, the classification process is more simple, with the fitness function considering only the settling time (for a given settling accuracy). Moreover, intuitively, a simplified fitness function helps the optimization algorithm converge faster to a better final solution.

Chapter 5
Design of a 7-bit 1 GS/s CMOS Two-Way Interleaved Pipeline ADC

Abstract This chapter discusses the design of a 7-bit 1 GS/s time-interleaved pipeline ADC, implemented in a standard digital 0.13 μm CMOS technology. All the sub-blocks used in the design of the ADC are also described in this chapter. Furthermore, specifications for the design and the architecture of the ADC are presented. One of the main objectives of this work was to integrate the proposed current-mode reference shifting (CMRS) MDAC circuit (Sect. 3.2) and verify its functionality in a working ADC structure. As secondary objectives, the proposed amplifier and flash quantizer (Chap. 4) circuits are also to be integrated in this prototype, in order to verify their performance, functionality, and usefulness in a pipeline ADC environment. Notice that, with the integration of both the proposed CMRS-MDAC and flash quantizer, the designed ADC precludes reference voltage circuitry, such as voltage buffers, decoupling capacitors, and damping resistors.

5.1 Specifications

The designed ADC prototype was not specified for any particular application. Nevertheless, specifications for the design were given and consisted mainly in proving the functionality of all the proposed circuits described in Chaps. 3 and 4, showing their potential in achieving high energy-efficiency in the design of MDAC-based ADCs. The target specifications for the ADC's implementation are given in Table 5.1.

5.2 Architecture

As mentioned in Table 5.1, a pipeline architecture has been selected to implement the ADC. To relieve the speed requirements of each building block, a two-channel time interleaved architecture is used. Although the time-interleaved structure itself

M. Figueiredo et al., *Reference-Free CMOS Pipeline Analog-to-Digital Converters*, Analog Circuits and Signal Processing, DOI: 10.1007/978-1-4614-3467-2_5,

Table 5.1 Targeted specifications for the designed ADC

Parameter	Target
Architecture	Pipeline
Structure	Fully differential
Technology	0.13 μm Single-Poly 8-Metal HS RFCMOS Process
Supply Voltage	1.2 V
Reference	0.5 V (differential)
Signal Bandwidth	Nyquist
Input Signal Full-Scale Range	1 Vpp
Input Signal CM Voltage	0.55 V
Resolution	7-bits
Sampling Rate	1 GS/s
ENOB @ Nyquist	>5 bits
INL	±0.5 LSB
DNL	±0.5 LSB

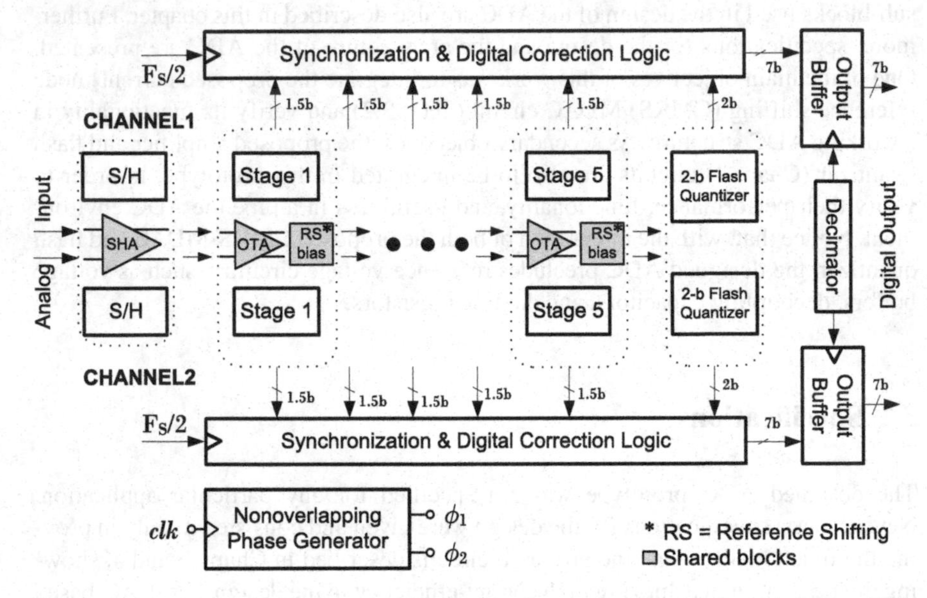

Fig. 5.1 Block diagram of the selected architecture for the two-way time-interleaved pipeline ADC

brings other difficulties to the design of the ADC, its benefits outweigh its drawbacks. A simplified block diagram of the overall architecture of the implemented ADC is shown in Fig. 5.1.

Concerning the analog part of the system, it is composed of a distributed front-end sample-and-hold (S/H), followed by five 1.5-bit stages and final 2-bit stage. The S/H is distributed to relax its specifications. Although time-interleaved timing mismatch

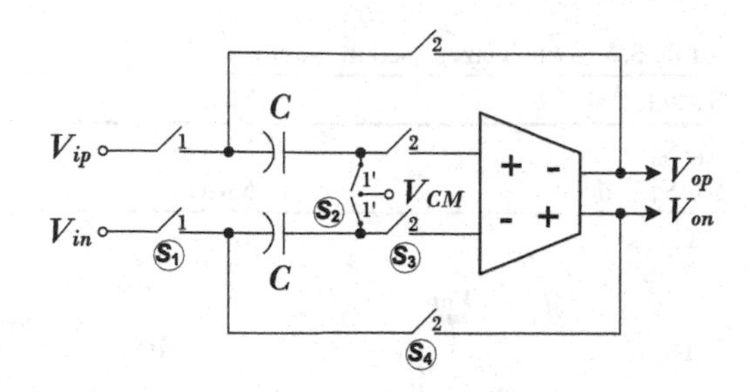

Fig. 5.2 Flip-around front-end sample-and-hold circuit

issues would be solved with a single S/H, designing one for 1 GS/s with Nyquist bandwidth would consume a lot of power and would make the timing scheme more complex. The S/H amplifier (SHA) is shared between channels to reduce power consumption. Each 1.5-bit stage employs the proposed CMRS-MDAC and 1.5-bit flash quantizer. To reduce power consumption, the amplifier (based on the design of the one proposed in Sect. 4.2), and reference shifting biasing circuit are shared between channels. The final stage is composed of a 2-bit flash quantizer which is based on the proposed architecture of Sect. 4.1. An on-chip analog buffer is used to buffer the common-mode voltage (V_{CM}) of the ADC.

Regarding the digital part of the system, the stages' output bits are synchronized and digital correction logic is used to correct for any possible errors. At the output, a decimator is used to reduce the rate at which the bits are outputted (given the high sampling rate) and finally, output buffers hold and drive the output bits. A nonoverlapping phase generator is used to produce the two main nonoverlapping clock signals with accurate phase shift.

5.3 Implementation Details

5.3.1 Sample-and-Hold

The architecture of the front-end sample-and-hold (S/H), as shown in Fig. 5.2, is based on the flip-around structure [181]. Fig. 5.2 depicts the fully differential circuit with multiplexer switches (found at the amplifier's input, S_3), which are used to switch the shared amplifier between channels. During the sampling phase, the input voltage is sampled on the capacitors. During the hold phase, the capacitors are connected between the input and output of the amplifier, and the output voltage becomes the sampled input voltage minus an error, mostly due to finite gain and GBW of the opamp. All switches are made of small NMOS transistors (sizes given in Table 5.2) controlled by a clock-bootstrapping scheme, which allows achieving the necessary early phases (indicated in Fig. 5.2 by 1') for signal-independent sampling. This scheme is discussed further on in this chapter.

Table 5.2 Switch sizes used in the S/H stage

Switch	Type	Size W/L (μm/μm)
S_1, S_3	NMOS	6/0.12
S_2, S_4	NMOS	3/0.12

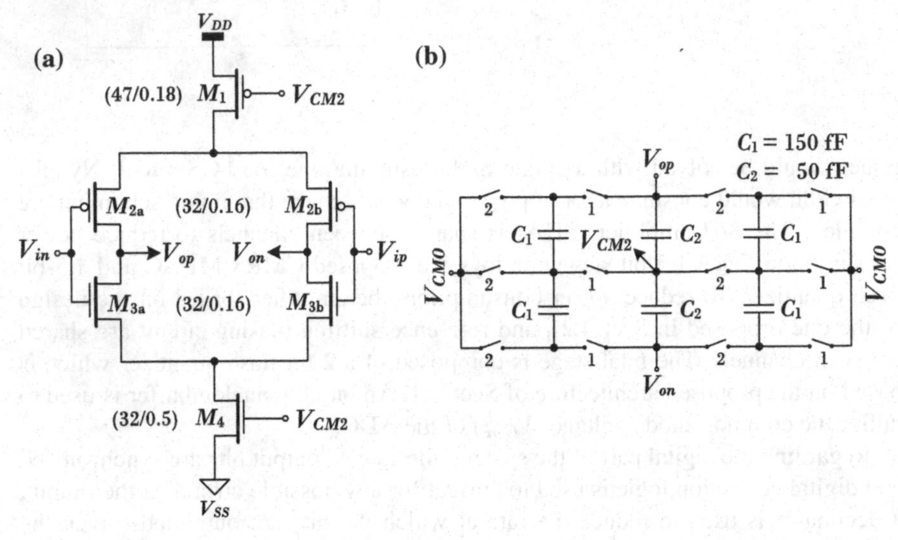

Fig. 5.3 Sample-and-hold amplifier: **a** Circuit. **b** Time-interleaved SC-CMFB circuit. Transistor sizes are given in μm

The capacitance value of the flip-around capacitors is 400 fF each. This value has been chosen in order to set the ADC's thermal noise power below half the ADC's quantization noise power. The amplifier used in the S/H is a single-stage inverter-input self-biased OTA, as depicted in Fig. 5.3a. Basically it is a single-stage version of the amplifier described in Sect. 4.2. It is identical to the output stage of that amplifier, in the sense that it has the same architecture, and the circuit is biased and the output CM level is controlled by the common-mode voltage (V_{CM2}) generated by the dedicated SC-CMFB circuit. This time-interleaved SC-CMFB circuit has a fixed load (C_2) and switched load (C_1) as shown in Fig. 5.3b. The sizes of the transistors and capacitance values are given in Fig. 5.3. Figure 5.4 depicts the open-loop Bode diagrams of the amplifier over PVT corners.[1] Table 5.3 summarizes the simulated performance of this amplifier for the typical corner.

[1] Corners considered: *tt*, *ss*, and *ff* for process; 1.2 V±5 % for supply voltage, and −40°, 27°, and 85° for temperature.

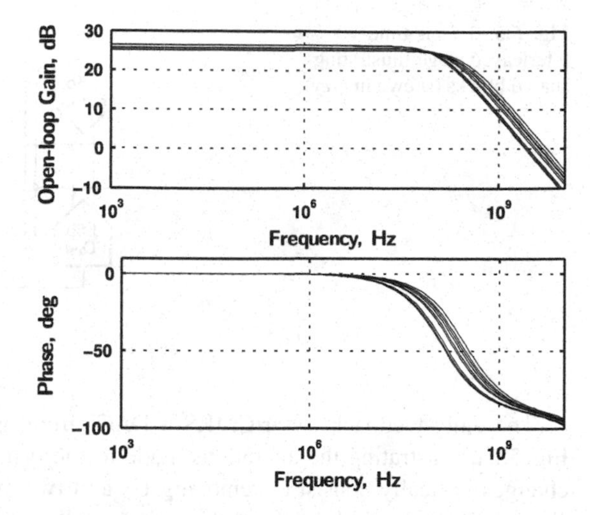

Fig. 5.4 Bode diagrams of the S/H amplifier over PVT corners

Table 5.3 Key performance summary of the S/H amplifier

Parameter	Value
Open-loop DC gain	26 dB
GBW[a]	3.7 GHz
Settling time[a] (0.1 V step)	249 ps
Settling time[a] (1 V step)	314 ps
Phase Margin[a]	90 °
Output Swing	1.3 V
Slew Rate	5400 V/μs
Total Input Offset	2.8 mV
Power @ 1.2 V	986 μW

[a] $C_L = 400$ fF

5.3.2 CMRS Multiplying-DAC

The MDAC architecture adopted for all stages of the implemented pipeline ADC was the 1.5-bit CMRS-MDAC presented and previously discussed in Sect. 3.2. Given that this MDAC circuit has already been fully detailed before, it will be the objective of this section to outline the design of the implemented 1.5-bit pipeline stage, discussing its time-interleaved operation with shared amplifier and shared reference shifting biasing (RS Bias) circuit, and indicating the alterations made to the original MDAC circuit. The amplifier, CMFB, and flash quantizer used in the stage will be discussed further on in this chapter.

The simplified block diagram of the implemented pipelined stage is shown in Fig. 5.5. The only blocks of interest here are the RS Bias and MDAC blocks. Figure 5.5 shows (shaded in grey) the blocks shared between the channels, which are the RS Bias and the OTA. First the MDAC then the RS Bias circuit will be discussed.

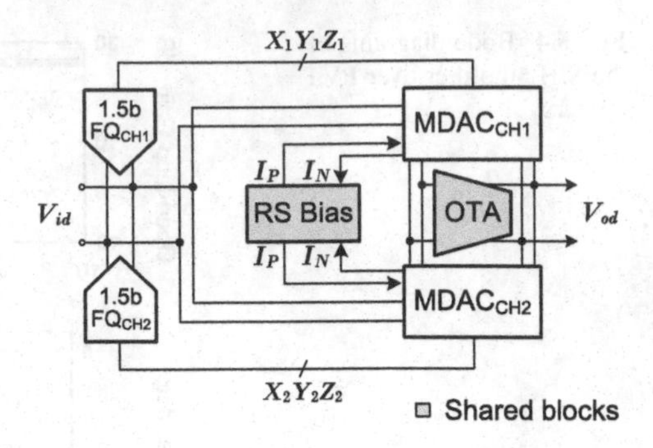

Fig. 5.5 1.5-bit time-interleaved stage illustrating shared blocks (shown in *grey*)

The unity feedback factor CMRS-MDAC circuit is repeated for convenience in Fig. 5.6a, illustrating the alterations made to the original circuit (Fig. 3.5a). These changes basically consist of removing C_3 and two switches, and adding switches S_4 and S_8. It should be noticed that two switches still remained from the original compensation circuit, as they were sufficient for parasitic compensation (using C_{p61}) at the 7-bit level. This type of compensation is sufficient up to 8-bit accuracy, whereas for 9-bit accuracy (and above), C_3 and all switches (with adequate matching) should to be used. Switch S_4 was added as a multiplexer given that the amplifier is shared between channels and S_8 was used to short-circuit the amplifier's inputs, destroying any residual charge, thus minimizing the memory effect due to the amplifier's input capacitance. ϕ_r is a very short pulse that is only high between the ending of the sampling phase and the start of the residue amplification phase (see Fig. 5.6b for an idea). Parasitic compensation was initiated at the design level by balancing switch sizes, but careful layout and iterative parasitic extraction ultimately dictated the process in which parasitic balancing was achieved. It is important to mention that this process, and therefore, the parasitic-aware layout of this MDAC circuit became quite intensive.

A timing diagram and the waveform of the output voltage of the MDAC are shown in Fig. 5.6b, c, respectively. After signal independent sampling takes place, the main capacitors of the MDAC are associated in series for the gain of two. This occurs, until the flash quantizer has made its decision, and X, Y, or Z is fed to the MDAC. In the case of X or Z, reference shifting occurs, which is illustrated in Fig. 5.6c for both modes. If too much time is given to the quantizer, the output of the amplifier could increase to beyond its output swing (because of the gain of two), driving its transistors into the triode region causing distortion. In the implemented ADC the quantizer's decision time was minimized, which helps the amplifier stay in saturation and also less current is needed for reference shifting because more time is available. From Fig. 5.6b, the residue amplification phase starts when ϕ_2 rises and ends when the residue is sampled by the following pipelined stage (ϕ_1'). Therefore, in terms of output voltage waveform, the output can be exponential if the stage is in Y mode or it can be a ramped signal in the other operation modes (shown in Fig. 5.6c).

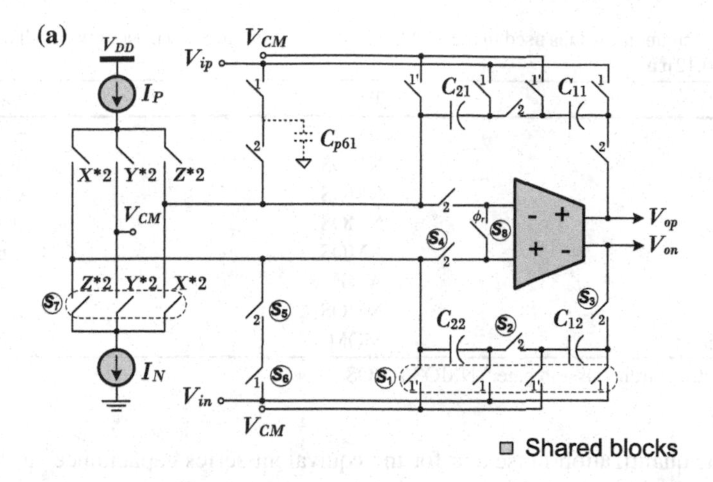

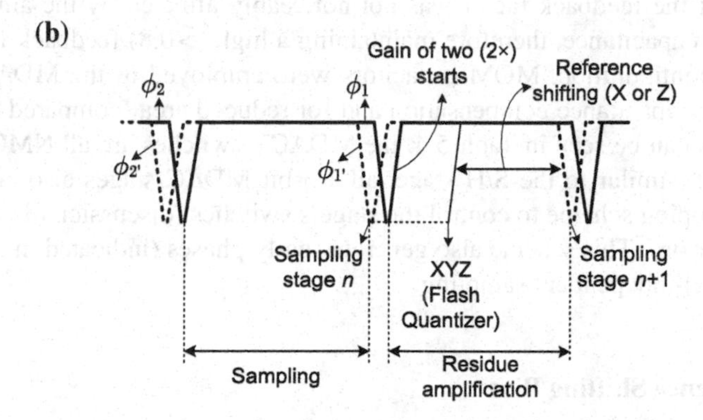

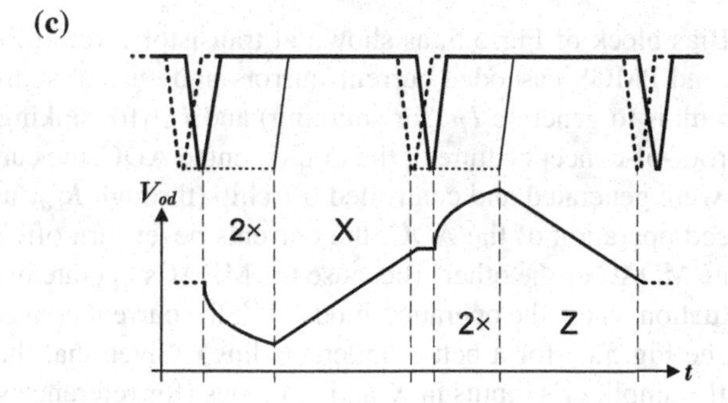

Fig. 5.6 Unity feedback factor CMRS-MDAC: **a** Circuit diagram. **b** Timing diagram of operations. **c** Output waveform of the opamp in X and Z modes

The dimensions of all the switches and capacitance values for the implemented MDAC are given in Table 5.4 (see Fig. 5.6a for parameter names). The capacitance value of 150 fF was chosen for two reasons: to guarantee that the thermal noise was

Table 5.4 Parameter values used in the MDACs. Switch widths are in μm (lengths of all transistors are set to 0.12 μm)

Device	Type	Value
S_1	NMOS	5
S_2, S_3	NMOS	1
S_4	NMOS	0.64
S_5	NMOS	0.32
S_6	NMOS	0.64
S_7	ATG[a]	0.64/1.15[b]
S_8	NMOS	2
All capacitors	MOM	150 fF

[a] Asymmetrical transmission gate [b] NMOS/PMOS

below the quantization noise and for the equivalent series capacitance[2] to be large enough so that the feedback factor was not noticeably affected by the amplifier's input parasitic capacitance, therefore maintaining a high (>0.8) feedback factor of the MDAC's configuration. MOM capacitors were employed in the MDAC for a better parasitic capacitance compensation and for reduced area (compared to MIM capacitors). As can be seen in Table 5.4, the MDAC's switches are all NMOS transistors because, similar to the S/H stage, all 1.5-bit MDAC stages also employ a clock-bootstrapping scheme to control the stage's switches (discussion of this topic is given further on). This scheme also generates early phases (indicated in Fig. 5.6a by 1') for signal-independent sampling.

5.3.2.1 Reference Shifting Bias

Regarding the RS Bias block of Fig. 5.5, as shown at transistor level in Fig. 5.7, it consists of NMOS and PMOS cascoded current mirrors and current sources. The objective of this circuit is to generate I_P (for sourcing) and I_N (for sinking). Given the prototype and proof-of-concept nature of the implemented ADC, the currents for the current mirrors were generated and controlled off-chip (through R_{cas} and R_b)[3]. Due to the high speed operation of the ADC, the currents never turn off. They are always steered to one MDAC or the other[4] (because the MDACs operate in opposite phases), or in the situation where the operation mode is Y, the current is steered from I_P directly to I_N (see Fig. 5.6a for a better understanding). Given that the current sources connect to the amplifier's inputs in X and Z modes (for reference shifting),

[2] Occurs during the amplification phase, where C_{1j} and C_{2j}, $j = \{1, 2\}$ are associated in series, and the equivalent capacitance becomes half the unit capacitance (150 fF), i.e., becomes 75 fF.

[3] Notice that, in a future design of this ADC, as an intellectual property (IP) product, these two currents would be generated on-chip either using an SC reference current generator [165] or using a replica MDAC together with a servo-loop (see Appendix A).

[4] The RS Bias circuit is shared between channels to reduce power consumption and circuit overhead, and therefore provides current for reference shifting for both MDACs.

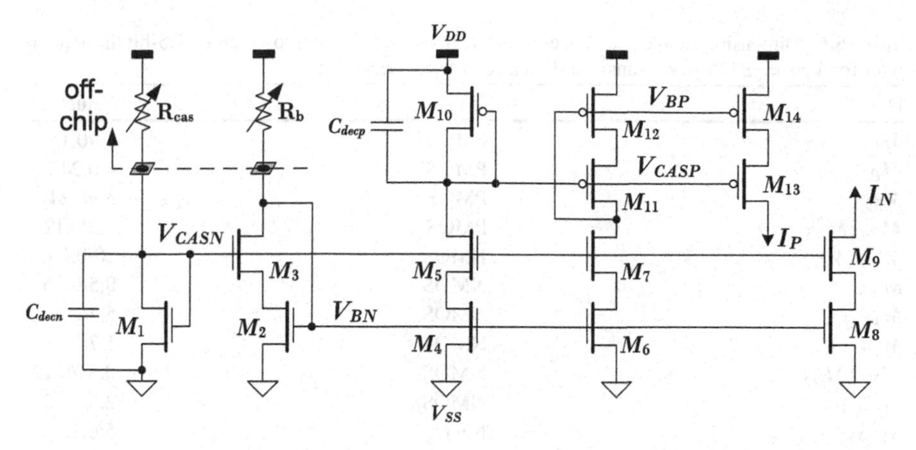

Fig. 5.7 Reference shifting biasing circuit, including current mirrors and current sources

Table 5.5 Transistor dimensions of the RS Bias circuit

Transistor	Size W/L (μm/μm)
M_1	3/1
M_2, M_4, M_6, M_8	7.92/0.2
M_3, M_5, M_7, M_9	16/0.17
M_{10}	3/0.3
M_{11}, M_{13}	16/0.12
M_{12}, M_{14}	7.9/0.16
Decoupling capacitors (MOS capacitors)	
C_{decn}, C_{decp}	200 fF

it is important that when the MDAC is in Y mode, the node between I_P and I_N maintains a voltage close to the amplifier's input common-mode voltage. This is crucial to reduce recovery time when the MDAC goes into X or Z mode again. To help minimize this problem, V_{CM} is connected between the current sources (in Y mode), therefore avoiding this node from floating and possibly drifting (refer to Fig. 5.6a).

The dimensions of the transistors used in the RS Bias circuit are given in Table 5.5.

5.3.3 Flash Quantizer

5.3.3.1 1.5-bit Flash Quantizer

The flash quantizer architecture adopted for all 1.5-bit stages of the implemented ADC is the one proposed and discussed in Sect. 4.1. Fig. 5.8 shows its circuit diagram,

Table 5.6 Dimensions and capacitances of the devices used in the implemented 1.5-bit flash quantizer for $V_{TH} = \pm 125$ mV. Transistor dimensions are given in μm

Device	Type	Value
M_{P1}, M_{P2}	PMOS	13/0.15
M_{P3}	PMOS	4/0.24
M_5	PMOS	30/0.24
M_{P4}, M_{P5}	PMOS	1.2/0.12
M_{N1}, M_{N2}	NMOS	10/0.16
M_{N3}	NMOS	8.5/0.16
M_{4A}	NMOS	5.5/2
M_{4B}	NMOS	1.7/2
M_{N4}, M_{N5}	NMOS	0.32/0.12
S_1, S_2, S_3	NMOS	2.5/0.12
S_4, S_5	NMOS	5/0.12
C_S	MOM	50 fF
C_{decn}, C_{decp}	MOS	50 fF

which includes the sampling circuit, the D-FFs, the XYZ encoder, and the digital logic for timing control of the various blocks. Recall, from Fig. 5.5, that each channel has its own flash quantizer, no sharing of any part of this circuit is done.

The analog part of the flash quantizer with sampling circuit, which is responsible for comparison, is shown in Fig. 5.8a. During the sampling phase (ϕ_1), the input signal is sampled on the C_S capacitors (signal-independent sampling occurs) and, simultaneously, the input signal is connected to the quantizer's inputs, allowing it to initiate comparison. During the comparison/decision phase (ϕ_2), only the held signal is connected to the quantizer's inputs, while the common-mode (CM) voltage of the input signal (node between capacitors) is connected to the quantizer's V_{CM} input. Devices S_4, S_5, C_{decn} and C_{decp} (MOS capacitors) have been added to hold the sampled signal's CM voltage more constant for the comparison/decision phase. This phase is divided into various sub-intervals (see ϕ_2 of Fig. 5.8d). First a certain amount of time is allocated for the quantizer to make a decision, after which the D-FFs open and the output of the quantizer is held. The circuit adopted for the D-FFs is shown in Fig. 5.8b. The second sub-interval is for the XYZ encoder (Fig. 5.8c), which takes the D-FF's outputs and clock signals to make a final decision on whether X, Y, or Z is high. This decision provides the MDAC with valid X, Y, or Z signals for reference shifting. Therefore, these signals are at a low state during the sampling phase, because the MDAC of the opposite channel will be in its residue amplification phase and could be using the RS Bias circuit (recall that the RS Bias circuit is shared between channels). The timing control circuit and complete timing diagram of operations are given in Fig. 5.8d.

The dimensions of the transistors, switches, and capacitances used in the analog part of the flash quantizer are given in Table 5.6 for a threshold (switching) voltage of ±125 mV (a full-scale input voltage of 1 Vpp has been specified).

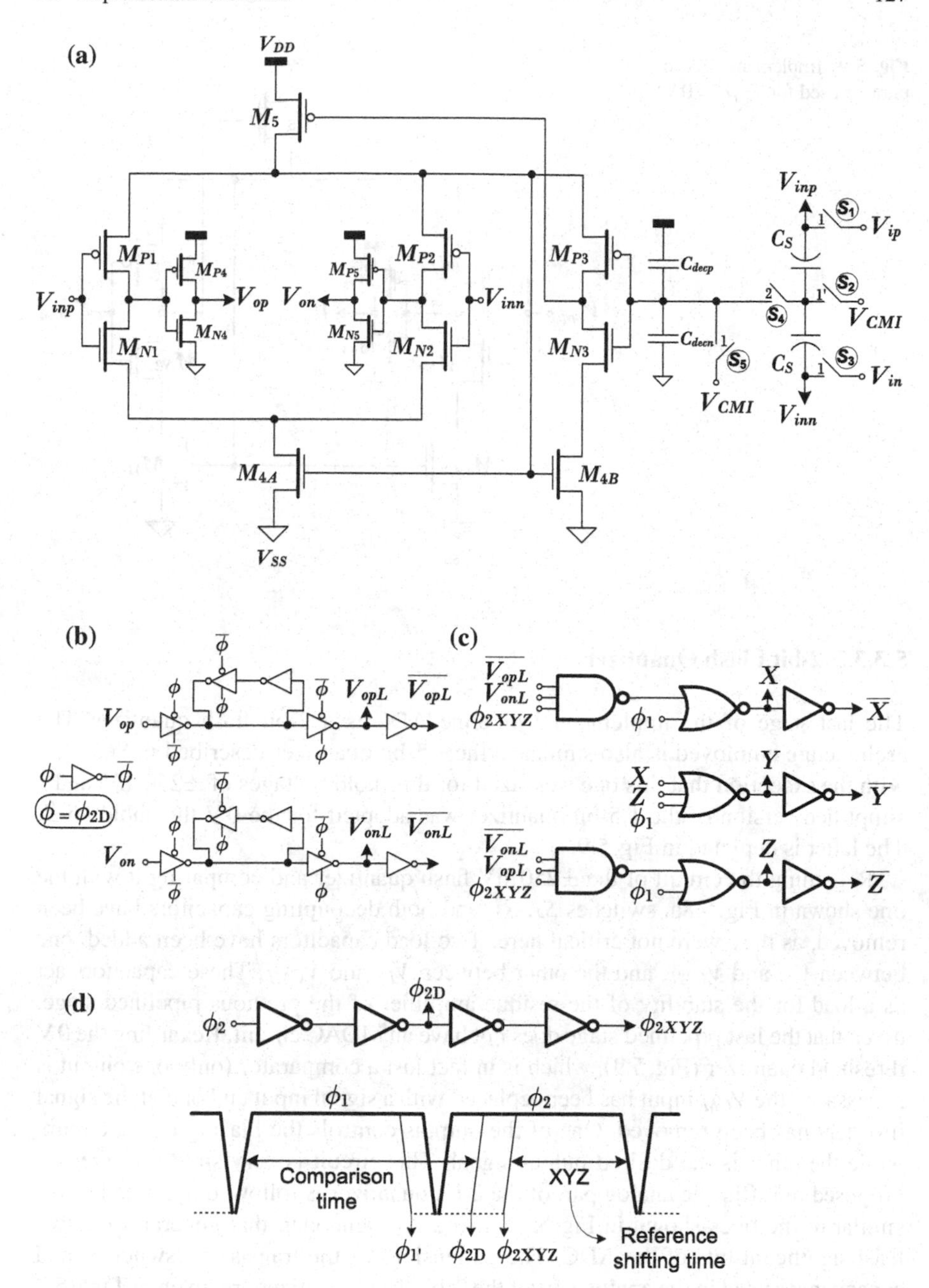

Fig. 5.8 Implemented 1.5-bit flash quantizer: **a** Circuit diagram of analog part with sampling circuit. **b** D-type flip-flops. **c** XYZ encoder. **d** Timing control and diagram of operations

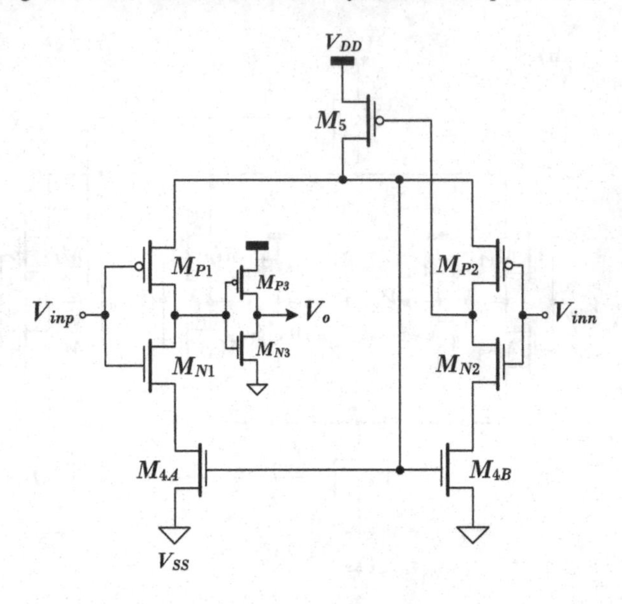

Fig. 5.9 Implemented comparator used for $V_{TH} = 0\,\mathrm{V}$

5.3.3.2 2-bit Flash Quantizer

The last stage of the implemented pipeline ADC is a 2-bit flash quantizer. The architecture employed is also similar to the 1.5-bit quantizer described in Sect. 4.1, with the exception that this one was sized for threshold voltages of $\pm 250\,\mathrm{mV}$, and a simplified version of the 1.5-bit quantizer was adopted for the 0 V threshold level. The latter is depicted in Fig. 5.9.

Regarding the circuit of the $\pm 250\,\mathrm{mV}$ flash quantizer and comparing it with the one shown in Fig. 5.8a, switches S_4, S_5, and both decoupling capacitors have been removed, as they were not critical here. Two load capacitors have been added, one between V_{ip} and V_{CMI}, and the other between V_{in} and V_{CMI}. These capacitors act as a load for the stability of the residue amplifier of the previous pipelined stage, given that the last pipelined stage does not have an MDAC circuit. Regarding the 0 V threshold quantizer (Fig. 5.9), which is in fact just a comparator (only one output is necessary), the V_{CM} input has been replaced with a signal input and one of the signal inverters has been removed. One of the outputs controls the biasing of the circuit, while the other is the desired output signal. This circuit is very similar to the one proposed in [10]. The analog part of the 2-bit quantizer is followed by three D-FFs, similar to the ones shown in Fig. 5.8b, and a 2-bit encoder, that generates the two least significant bits of the ADC. The dimensions of the transistors, switches, and capacitances used in the analog part of the 2-bit flash quantizer are given in Table 5.7 for the threshold levels of $\pm 250\,\mathrm{mV}$ (see Fig. 5.8a for component names) and 0 V (Fig. 5.9).

Table 5.7 Dimensions and capacitances of the devices used in the implemented 2-bit flash quantizer for $V_{TH} = \pm 250\,\text{mV}$ and $V_{TH} = 0\,\text{V}$. Transistor dimensions are given in μm

Device	Type	Value
S_1, S_2, S_3	NMOS	5/0.12
C_S	MOM	100 fF
C_L	MOM	100 fF
$V_{TH} = \pm 250$ **mV**		
M_{P1}, M_{P2}	PMOS	9/0.12
M_{P3}	PMOS	60/0.12
M_5	PMOS	30/0.24
M_{P4}, M_{P5}	PMOS	0.58/0.12
M_{N1}, M_{N2}	NMOS	60/0.12
M_{N3}	NMOS	4/0.36
M_{4A}	NMOS	6.5/1
M_{4B}	NMOS	26/1
M_{N4}, M_{N5}	NMOS	0.16/0.12
$V_{TH} = 0$ **V**		
M_{P1}, M_{P2}	PMOS	10/0.14
M_5	PMOS	14/0.24
M_{P3}	PMOS	0.58/0.12
M_{N1}, M_{N2}	NMOS	10/0.14
M_{4A}	NMOS	3.3/1
M_{4B}	NMOS	3.5/1
M_{N3}	NMOS	0.16/0.12

5.3.4 Opamp and CMFB

The architecture of the amplifier adopted for all MDAC stages is identical to the one proposed and discussed in Sect. 4.2. This inverter-based self-biased amplifier was employed given its high speed at low power consumption, i.e., its high energy efficiency. Given that no structural changes were made to the original amplifier and CMFB circuits, no circuit diagram will be shown here. The sizes of the transistors and capacitance values used in the amplifier and CMFB circuits are given in Table 5.8. Refer to Fig. 4.12 and 4.13 for device names. The architecture used for SC-CMFB, CMFB_2, is identical to that used in the amplifier of the S/H stage (see Fig. 5.3b) given that this amplifier is also shared between channels.

Fig. 5.10 depicts the open-loop Bode diagrams of the amplifier over PVT corners. Table 5.9 summarizes the simulated performance of this amplifier for the typical corner.

Table 5.8 Transistor dimensions and capacitance values used in the implemented amplifier and CMFB circuits. Transistor dimensions are given in μm

Device	Type	Value
M_{11}	PMOS	7.94/0.28
M_{12a}, M_{12b}	PMOS	10.72/0.12
M_{13a}, M_{13b}	NMOS	8.56/0.12
M_{14a}, M_{14b}	NMOS	3/0.95
M_{21}	PMOS	71.44/0.38
M_{22a}, M_{22b}	PMOS	36.28/0.14
M_{23a}, M_{23b}	NMOS	20/0.15
M_{24}	NMOS	64/1.16
M_{31}	PMOS	16.82/0.14
M_{32}	NMOS	15.84/0.13
M_{33}	PMOS	17.92/0.13
M_{34}	NMOS	8.68/0.13
C_{Ca}, C_{Cb}	MOM	90 fF
C_{CM}	MOM	120 fF
CMFB$_2$		
C_1	MOM	90 fF
C_2	MOM	250 fF
All switches	NMOS	1/0.12

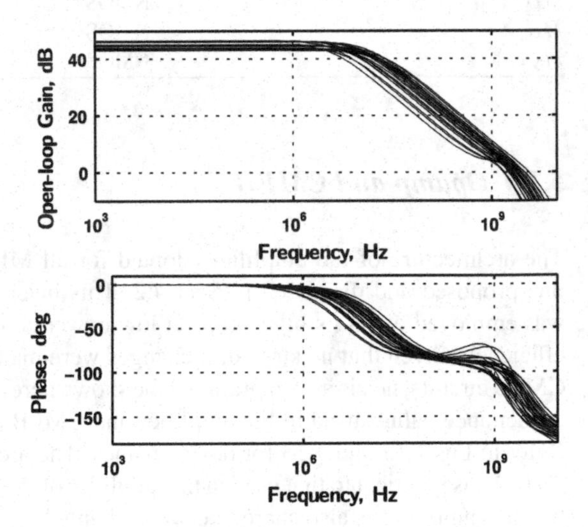

Fig. 5.10 Bode diagrams of the MDAC amplifier over PVT corners

5.3.5 Switches and Clock-Bootstrapping Circuits

As already mentioned before and shown in the tables that indicate information about the switches used in the various blocks discussed until now, most switches are simply NMOS transistors. In a pipelined stage, the only switches that are not NMOS-only

Table 5.9 Key performance summary of the amplifier used in the MDACs

Parameter	Value
Open-loop DC gain	45 dB
GBW[a]	2.3 GHz
Settling time[a] (0.1 V step)	511 ps
Settling time[a] (1 V step)	736 ps
Phase Margin[a]	57 °
Output Swing	1.4 V
Slew Rate	3400 V/μs
Total Input Offset	4.1 mV
Power @ 1.2 V	1.18 mW

[a] $C_L = 600\,\text{fF}$

are those used to steer the current for reference shifting in the MDAC circuits, which are ATGs (asymmetrical transmission gates) controlled by the X, Y, and Z phases. Using small NMOS-only switches is made possible by driving all switches with bootstrapped voltages. A block diagram of the scheme that makes this possible is shown in Fig. 5.11a, accompanied by a timing diagram given in Fig. 5.11b. The main clock generator (described in the next section) generates two nonoverlapping clock phases, ϕ_1 and ϕ_2. Both phases are connected to each stage, where they are regenerated and bootstrapped phases are produced to be used by all blocks in the stage. The block diagram of Fig. 5.11a only depicts the blocks used for one main phase. Therefore, each stage has two of these blocks. Circuit operation is as follows: the three inverters regenerate and delay the original main phase (ϕ). This phase and its inverted counterpart (ϕ_n) connect to one clock-bootstrapping circuit, which is responsible for creating the early phase, ϕ_{Be} ('B' for bootstrapped and 'e' for early). The remaining inverters are used to delay ϕ even more, creating ϕ_d and ϕ_{dn}. These connect to the other bootstrapping circuit, which creates ϕ_B. The delay between ϕ and ϕ_d is sufficient to create a delay between ϕ_{Be} and ϕ_B for signal-independent sampling. Given that only bootstrapped phases are used to drive all switches, the input of the clock-bootstrapping circuits could not be any specific signal voltage, it could only be a DC voltage. This voltage, V_{CB}, was set to 1 V, and was generated and controlled off-chip for testability purposes. The timing diagram of Fig. 5.11b shows the bootstrapped and early phases used in each pipelined stage.

The circuit adopted for both clock-bootstrapping blocks is shown in Fig. 5.12 [41]. Given that each bootstrapping circuit drives a different number of switches, hence a different capacitive load, they have different transistor sizings, which are given in Table 5.10. Figure 5.13 depicts the simulated result of each clock-bootstrapping circuit, illustrating the nonoverlapping, rise and fall times, and the noncritical overlap.

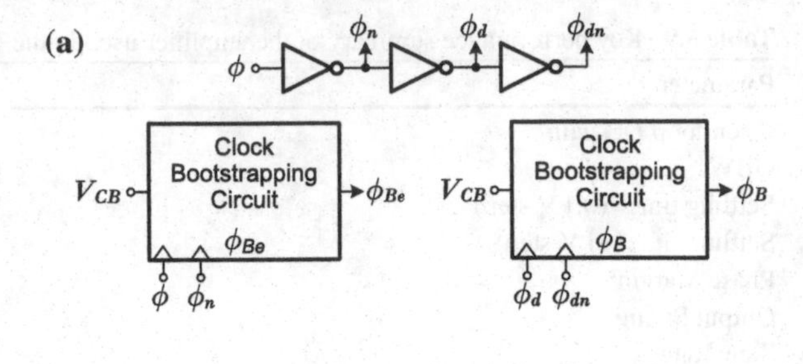

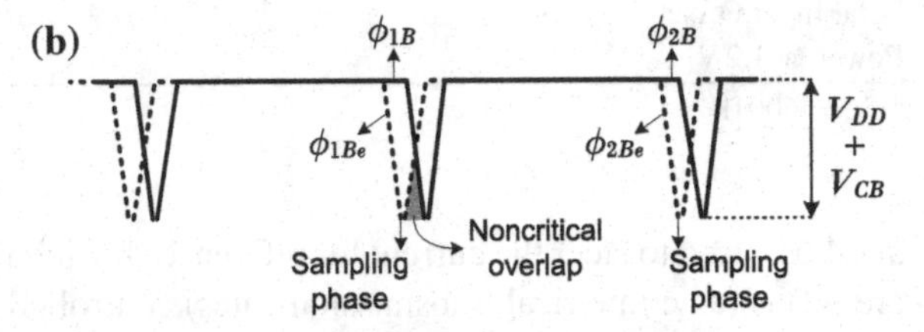

Fig. 5.11 Implemented clock-bootstrapping scheme: a Block diagram. b Timing diagram

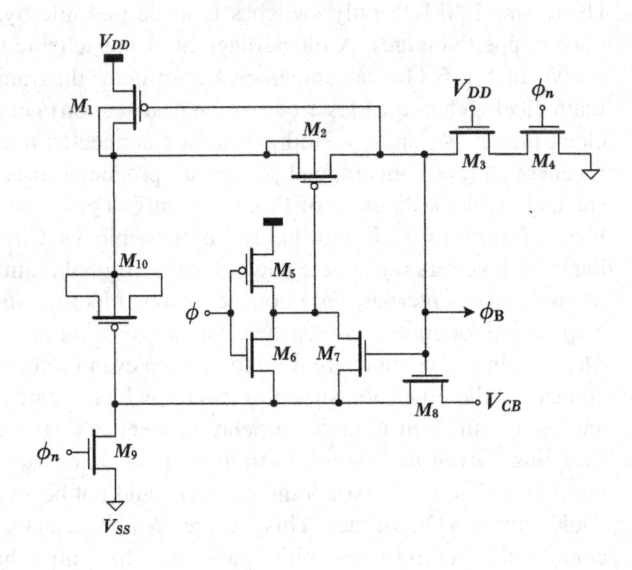

Fig. 5.12 Clock-bootstrapping circuit. PMOS and NMOS transistors with undefined bulk have their bulk connected to V_{DD} and V_{SS}, respectively

5.3.6 Clock Generator

The clock generator adopted for the ADC was the conventional nonoverlapping clock generator [105], which synthesises the two main phases, ϕ_1 and ϕ_2. The circuit diagram of the clock generator is shown in Fig. 5.14a. Each digital block of the clock generator was sized with the help of Monte Carlo simulations in order to reduce clock jitter and guarantee equal duty-cycle, given the high-speed and time-interleaved nature of the converter. Both main phases connect to each stage of the converter where they are buffered and bootstrapped phases are created as described in Sect. 5.3.5.

Table 5.10 Transistor dimensions used in the implemented clock-bootstrapping circuits. Dimensions are given in μm

Transistor	Size
Clock-bootstrapping circuit ϕ_B	
M_1, M_5	20/0.12
M_2	30/0.12
M_3, M_4, M_8	14/0.12
M_6, M_7	5/0.12
M_9	7/0.12
M_{10}	10/10
Clock-bootstrapping circuit ϕ_{Be}	
M_1, M_5	12/0.12
M_2	16/0.12
M_3, M_4, M_8	12/0.12
M_6, M_7	3/0.12
M_9	6/0.12
M_{10}	9/9

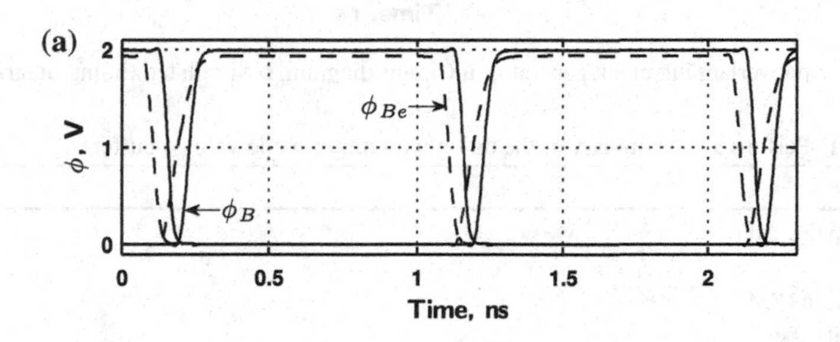

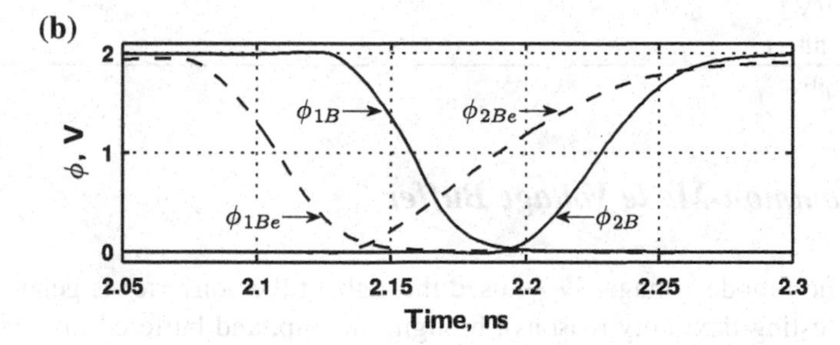

Fig. 5.13 **a** Simulated timing diagram of the clock-bootstrapping circuits' outputs. **b** Zoom of the timing diagram

Fig. 5.14b depicts the simulated result of the two synthesised phases of the clock generator. The rise and fall times are approximately 50 ps and the nonoverlapping time is 80 ps (measured at 0.1 V).

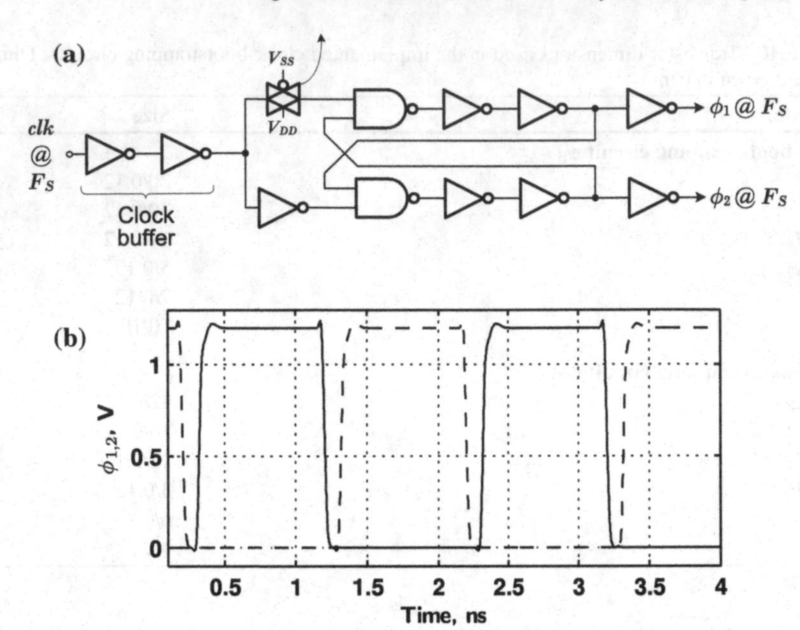

Fig. 5.14 Nonoverlapping clock generator: **a** Circuit diagram. **b** Simulated timing diagram

Table 5.11 Performance summary of the on-chip common-mode voltage buffer

Parameter	Value
Open-loop DC gain	20 dB
GBW[a]	25 MHz
Phase margin (PM)[a]	95 °
Total Input offset	9 mV
Power @ 1.2 V	1.5 mW
Fixed load (filter)	100 pF

[a] $C_L = 100\,\text{pF}$

5.3.7 Common-Mode Voltage Buffer

The common-mode voltage, V_{CM}, used throughout the converter is generated off-chip (for testing flexibility reasons), brought on-chip, and buffered internally. The circuit diagram of the amplifier used in the buffer is shown in Fig. 5.15, which represents a single stage cascode PMOS-input circuit. It was configured in a unity-gain configuration to buffer V_{CM}. In order to limit the amplifier's output noise and stabilize its output voltage (due to the periodic clocking), the amplifier was loaded with a decoupling capacitor composed of two MOS capacitors (one PMOS capacitor and one NMOS capacitor) with a total capacitance of, approximately, 100 pF.

Table 5.11 summarizes the electrically simulated performance of this amplifier.

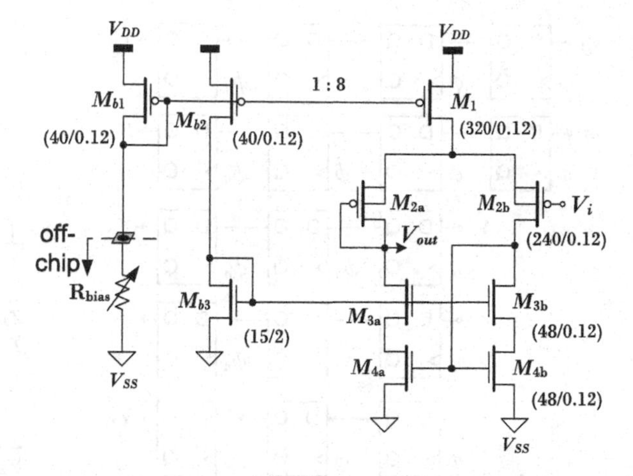

Fig. 5.15 Circuit diagram of the amplifier used to buffer V_{CM}

Notice that this buffer has been sized targeting the maximum conversion rate of 1 GS/s and its power dissipation does not scale with the clock frequency. Again, in a future redesign of this ADC as an IP, an on-chip SC current reference generator would also be integrated and R_{bias} would not be required.

5.3.8 Digital Backend

The digital backend of the converter is primarily composed of synchronization logic, digital correction logic, and digital output buffers. Given the high speed nature of the converter, two protective measures were adopted for acquiring the output bits. One measure employed two random access memories (RAMs), one for each channel, where the output bits are written to the RAM at full clock speed and then brought off-chip at a much lower frequency. The second measure consisted of a time-interleaved decimator, where the output buffer is clocked at a lower rate (decimated rate) and the bits are brought off-chip at this lower frequency. A digital input, SEL, was used to switch between both modes. The decimator will be discussed in the following section, and no more references to the RAM will be made, since it was not found necessary during testing. As already explained in the introductory chapter of this book, the digital outputs of each stage (Y and Z) need to be synchronized before digital correction. The synchronization logic block diagram is depicted in Fig. 5.16a. This circuit is composed of shift registers, which consist of D-FFs, and delay and buffering logic. All inputs connect to their respective D-FFs and the final outputs connect directly to the correction logic. The input clock is generated by the flash quantizer of the first MDAC stage (of the pipeline). This allows the synchronization logic to know when the bits, of each stage, are ready to be saved (clocked).

As mentioned, the output bits when time-aligned, are connected to the correction logic. This circuit is composed of a serial sequence of 1-bit full adders which produce

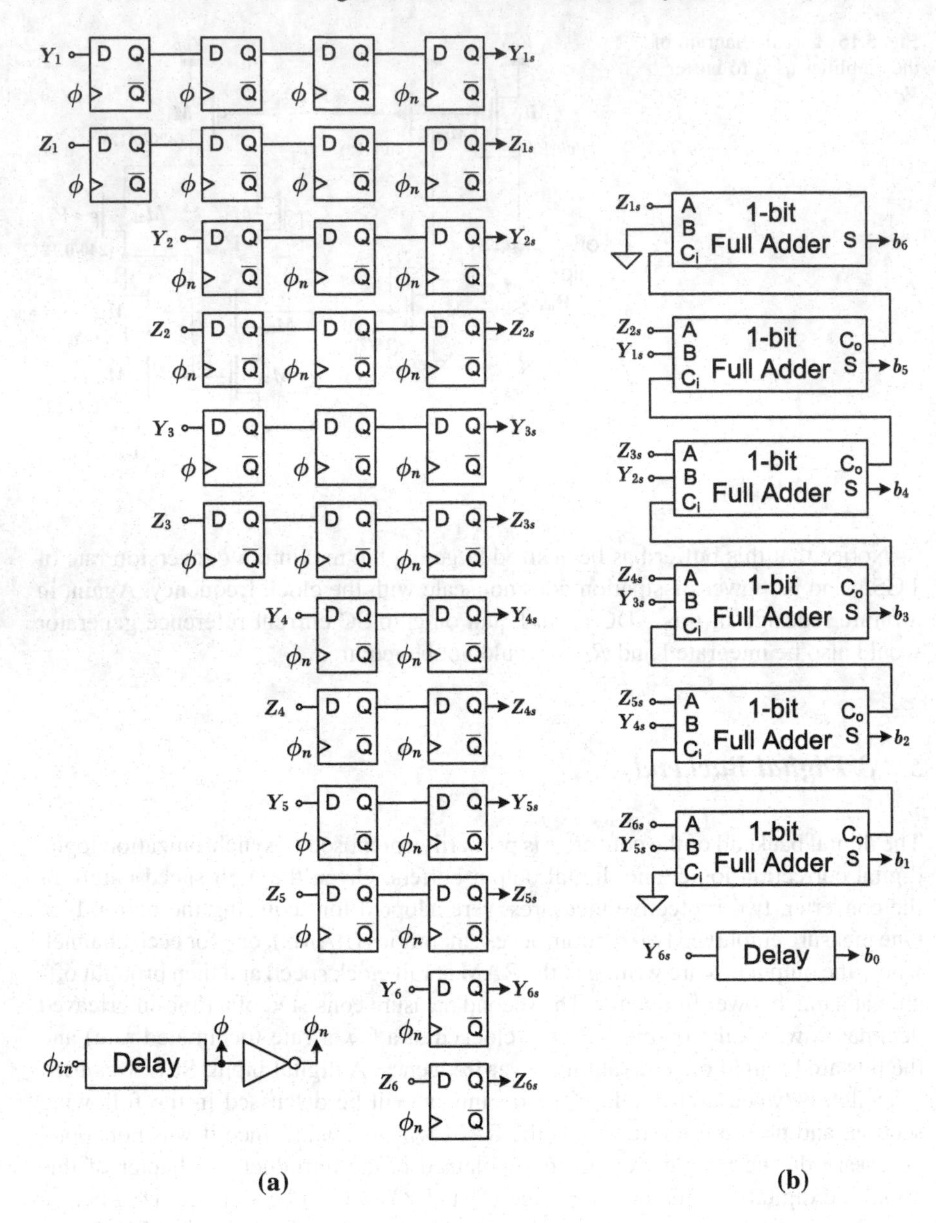

Fig. 5.16 Digital backend: **a** Synchronization. **b** Correction logic

the final (and corrected) digital outputs. The block diagram of this circuit is shown in Fig. 5.16b. Finally, the outputs from the correction logic are fed to the output buffer circuit. Here the bits are held using a bank of D-FFs followed by two cascaded inverters for buffering. The input clock to this circuit is generated and controlled by

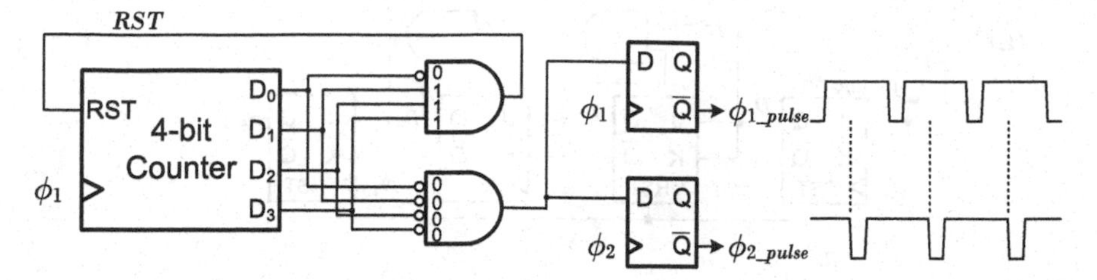

Fig. 5.17 Block diagram of the time-interleaved decimator

the decimator. If the decimator is selected to be used, a lower frequency clock is generated and fed to the digital output block. In this way the circuit is clocked at a lower frequency, hence the decimation effect. If the RAM is selected, the buffer circuit is clocked at full speed, and the bits are written to the RAM.

Due to time-interleaving, two of each of the above mentioned circuits, except for the decimator, had to be employed.

5.3.9 Decimator

The high speed nature of the implemented converter demanded for a decimator at the output to reduce the rate at which the bits are outputted, therefore facilitating data acquisition through external measurement equipments. This section describes the implemented time-interleaved decimator. The objective of this circuit is to discard samples but in a time-interleaved way, i.e., acquire a sample from one channel, then discard samples, then acquire a sample from the other channel, then discard samples, after which the process repeats itself. A time-interleaved decimator had to be implemented because using one conventional decimator [67] per channel, causes the equivalent to a timing mismatch error in the output FFT. In other words, when the outputs of each channel (separately decimated) are interleaved for post-processing, a timing error occurs, with a similar effect in the FFT to a timing mismatch error of time-interleaved ADCs.

The time-interleaved decimator, as described by its block diagram in Fig. 5.17, is composed of a 4-bit counter, and logic to control the reset and clock signals. The reset signal is used to restart the counter, while the clock signals (ϕ_{i_pulse}) are used to clock the digital output buffers with a lower frequency clock. In order to synchronize the output clock signals with the rest of the circuit's clocking, two D-FFs are used. The counter circuit diagram is shown in Fig. 5.18a, which is composed of four cascaded JK flip-flops (JK-FF). Besides counting, this circuit also serves to divide the main clock (ϕ_1 was used) frequency by 16, which is then sent off-chip to synchronize the logic analyzer for data capture (D_3 of Fig. 5.18a was used). The implemented converter has two channels, which means that a digital word is outputted every edge of ϕ_1, i.e., when ϕ_1 falls, a digital word from channel 1 is ready, when ϕ_1 rises,

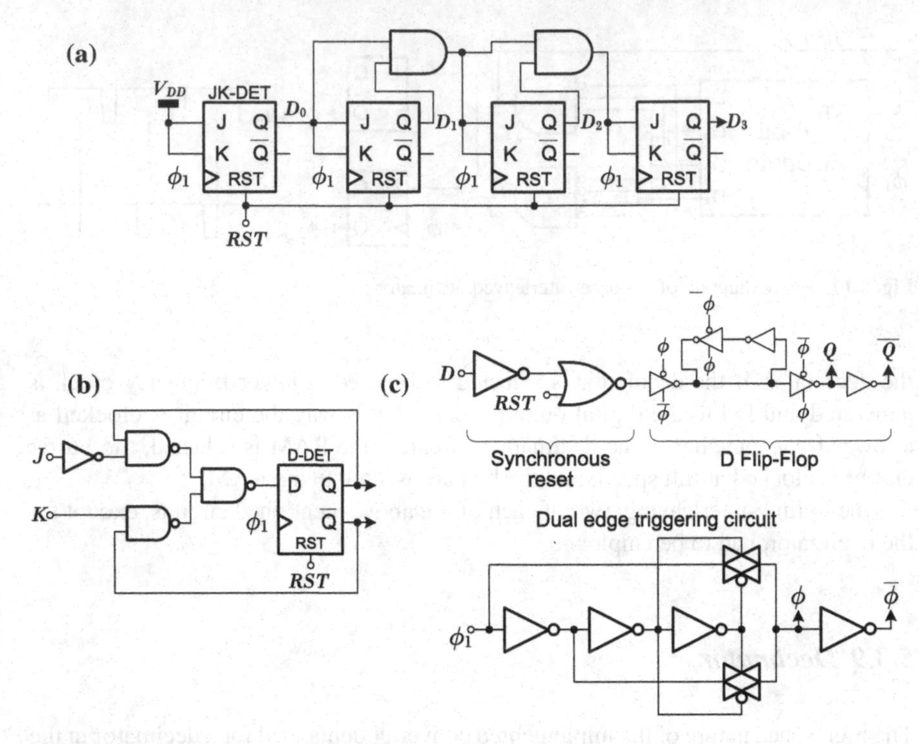

Fig. 5.18 Circuit diagrams of: **a** 4-bit counter. **b** JK-FF with synchronous reset. **c** D-FF with dual edge triggering circuit with synchronous reset

coinciding with ϕ_2 falling, a digital word from channel 2 is ready. Therefore, the JK-FFs need to count both falling and rising edges of ϕ_1, and to this end, dual edge triggered (DET) JK-FFs have been used, as depicted in Fig. 5.18b. In the composition of a JK-FF is a D-FF with synchronous reset, as shown in Fig. 5.18c. The circuit that allows the dual edge triggering is illustrated in Fig. 5.18c [116].

To sum up the decimator's operation: it counts 14 edges (rising and falling) of ϕ_1 and on the 15th clock edge, it signals the output buffer of channel 1 to hold the digital word at its input. The decimator is then reset and after 14 counts it signals the output buffer of channel 2 to hold the next digital word. The decimation factor is 15, which is an odd number because an even number would cause the decimator to constantly clock the same channel. Therefore, the decimator 'ping-pongs' between channels. When the output samples are time-interleaved during the post-processing stage (FFT, etc.), they will be in the correct order, and coherent sampling will not have been lost, nor timing errors will appear in the FFT.

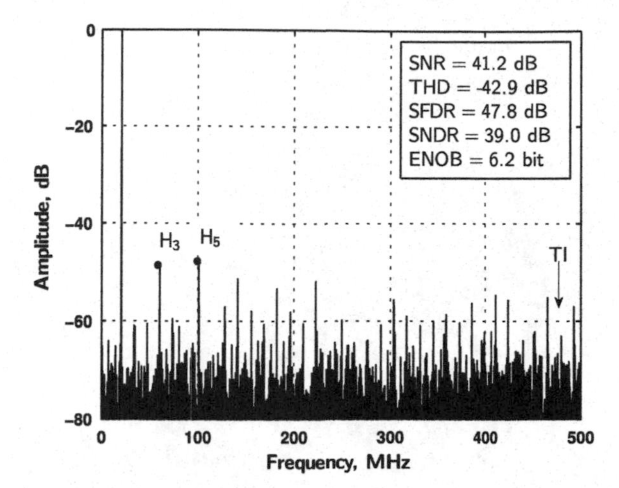

Fig. 5.19 4096-point FFT of the electrically simulated 7-bit 1 GS/s pipeline ADC for a 20 MHz input

5.3.10 Complete ADC

To finalize this chapter, the dynamic performance of the complete ADC is evaluated. An electrically simulated 4096-point FFT of the 7-bit ADC operating at 1 GS/s with a -1 dBFS 20 MHz input signal (for typical conditions) is shown in Fig. 5.19. The ADC obtains a SFDR of 47.8 dB and an SNDR of 39 dB, which leads to a 6.2 bit ENOB. At this sampling frequency, the total reference shifting current per stage is 68 μA and the total power consumption of the ADC is 19.5 mW, which results in a FoM of 265 fJ/conv.-step.

Chapter 6
Integrated Prototypes and Experimental Results

Abstract This chapter discusses the implementation of two integrated circuit (IC) prototypes with the objective of evaluating their performance and functionality. The first circuit concerns the two-stage inverter-based self-biased amplifier presented and described in Sect. 4.2 The second prototype concerns the 7-bit 1 GS/s two-way interleaved pipeline ADC, which was presented in Chap. 5. For each prototype, a floorplan and layout will be shown, and some design and layout considerations will be given. Equally, the designed printed circuit board (PCB) and the respective test setup used for each circuit evaluation will be presented. Each IC prototype section will end with the experimental results achieved and a comparison will be carried out with the state-of-the-art. Regarding the technology employed to design and implement the circuits, a standard digital 0.13 μm 1.2 V 1P-8M CMOS process has been used. No special device options such as, low- or high-V_T, twin- or triple-well or low leakage devices, etc., were used. For capacitors, the technology's Metal-Insulator-Metal (MIM) capacitors, MOS capacitors, and our design group's proprietary Metal-Oxide-Metal (MOM) capacitors were used.

6.1 Two-Stage Inverter-Based Self-Biased Amplifier

To validate the theoretical results derived in Sect. 4.2, an optimization using the technique also described in the referred section, was carried out. The DM settling time at 0.024 % error (12-bit accuracy) was set to 150 ns for a capacitive load of 4 pF. Due to thermal noise constraints related with the target application, the minimum value for C_C was set to 0.5 pF. No common-mode (CM) settings were defined, the OS was maximized, the power dissipation minimized, and the amplifier was optimized to provide stability for closed-loop gains higher than 2. The CM voltage of the input signal and V_{CMO} were set to 550 mV. The value of V_{CMO} is obtained by taking half the lowest supply voltage (considering ±10 % variations), i.e., $V_{CMO} \approx 0.9 \times V_{DD}/2$.

M. Figueiredo et al., *Reference-Free CMOS Pipeline Analog-to-Digital Converters*, Analog Circuits and Signal Processing, DOI: 10.1007/978-1-4614-3467-2_6,

Table 6.1 Transistor dimensions, and capacitor and resistor values used in the implemented amplifier and CMFB circuits. Transistor dimensions are given in μm

Device	Type	Value
M_{11}	PMOS	209/5.2
M_{12a}, M_{12b}	PMOS	225/1.6
M_{13a}, M_{13b}	NMOS	50.9/0.9
M_{14a}, M_{14b}	NMOS	22.8/0.3
M_{21}	PMOS	62.9/8.6
M_{22a}, M_{22b}	PMOS	80.4/0.7
M_{23a}, M_{23b}	NMOS	194/1.2
M_{24}	NMOS	12.6/1.7
M_{31}	PMOS	103/1.5
M_{32}	NMOS	5.4/1.4
M_{33}	PMOS	103/1.5
M_{34}	NMOS	5.4/1.4
C_{Ca}, C_{Cb}	MIM	510 fF
C_{CM}	MIM	510 fF
CMFB		
C_{CMFB}	MIM	100 fF
R_{CMFB}	Nonsalicide P+ Poly Resistor	50 kΩ

Table 6.1 summarizes the device sizing of the implemented amplifier (see Fig. 4.12 for device names). All capacitors were implemented using MIM capacitors and the resistors (for the CMFB circuit) were implemented with nonsalicide P+ poly resistors (readily available, as a standard option, from the technology).

6.1.1 Floorplan and Layout

Figure 6.1 shows the floorplan, the layout, and the chip photograph of the implemented amplifier. Figure 6.1c highlights the amplifier core, the on-chip continuous time CMFB circuit, the compensation capacitors C_C, and the CM stabilization capacitor C_{CM} (half on each side of the amplifier). The total area occupied by the amplifier, with compensation capacitors and CMFB, is $179 \times 69\,\mu m^2$. To facilitate the measurement stage of our design a continuous time CMFB circuit was implemented. This on-chip CMFB substitutes the SC-CMFB$_2$ circuit of Fig. 4.13a, but has the same purpose and consists of a 100 fF capacitor in parallel with a 50 kΩ resistor loading each output, connected at a common voltage, which represents V_{CM2} (see Fig. 6.3a for an idea). Due to this resistance value, the CMFB circuit slightly reduced the amplifier's DC gain.

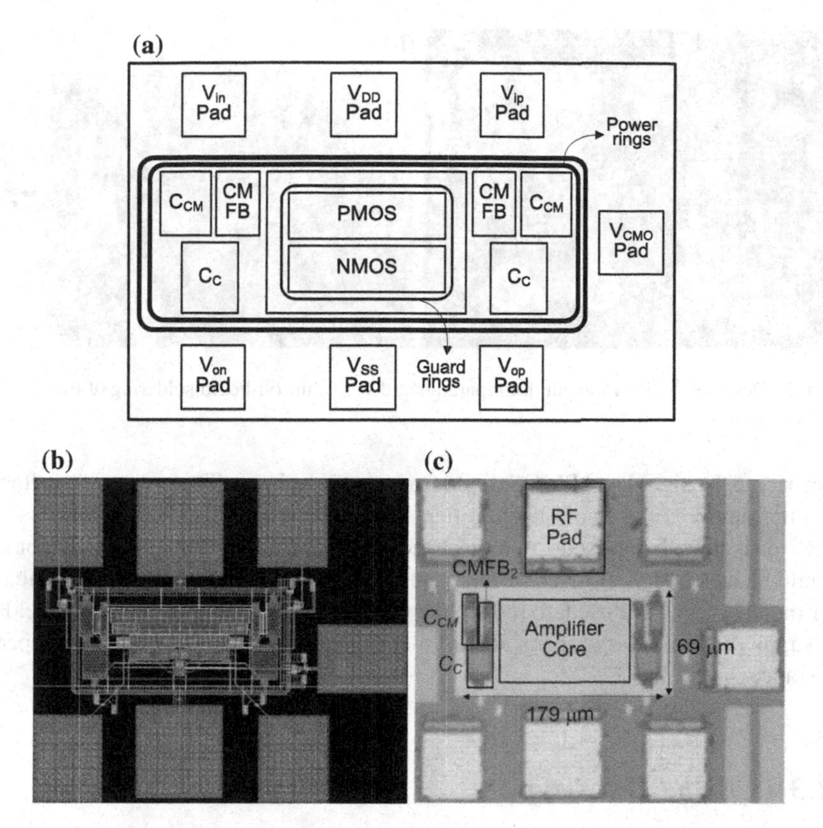

Fig. 6.1 Implementation of the two-stage inverter-based self-biased amplifier: **a** Floorplan. **b** Layout. **c** Chip photograph

6.1.2 Design Considerations

It will be the objective here to briefly describe some considerations taken during the design and layout of the amplifier. As can be seen in the amplifier's floorplan, its core was separated from the rest of the circuit by means of an n-well guard ring. This is not mandatory because the circuit is entirely analog, nevertheless it is good practice.

Given that the input signals, V_{ip} and V_{in}, and the common-mode voltage, V_{CMO}, are connected directly to the gates of the transistors, special precautions had to be taken. This was necessary to avoid electrostatic discharge (ESD) issues. These discharges can seriously damage transistors, leaving them useless or with erroneous functionality. To protect the amplifier, two cascaded ESD protection circuits were used. The primary circuit was composed of two RF ESD diodes, one connected from V_{SS} to the signal voltage and the other from the signal to V_{DD}. The secondary ESD protection consisted of a $100\,\Omega$ resistor in series with the signal and another set of diodes, but these were implemented using diode connected transistors (NMOS diode

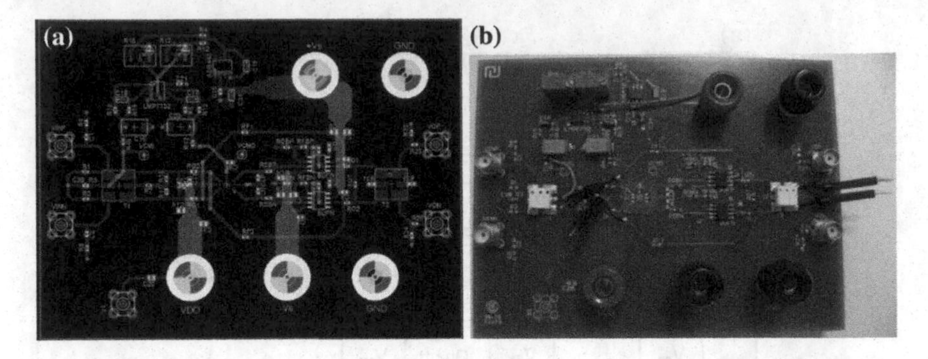

Fig. 6.2 Designed PCB: **a** Layout. **b** Prepared board with chip-on-board soldering of the die

from V_{SS} to the signal and PMOS diode from signal to V_{DD}). The protection voltages were the supply voltages of the amplifier, V_{DD} and V_{SS}.

RF pads have been used to connect the die to the PCB. The pads used for the signal voltages and V_{CMO} were implemented with a minimum size and only the top four metal layers were used, to reduce the parasitic capacitance to the substrate. For the supply voltages, larger pads and all metal layers were used to reduce the pad's resistance.

6.1.3 PCB and Test Setup

In order to evaluate the performance of the implemented amplifier, a PCB was designed using EAGLE layout editor. The PCB was designed with only two layers: top layer for routing and bottom layer for routing and ground plane. The on-board components used consisted of a voltage reference (ADR440), two low noise opamps for driving the CM voltages (LMP7732), two 1:1 low frequency wideband transformers for single-to-differential conversion (TTWB3010), and two high speed buffers to buffer the amplifier's outputs (AD8000). Given that the circuit-under-test is a transconductance amplifier and resistive feedback was to be employed for closed-loop testing, buffers had to be used for the circuit to maintain its DC gain. This situation introduced a new issue, which was the stability of the complete closed-loop circuit composed of implemented amplifier and buffers. To maintain stability and not affect the measurements of the proposed amplifier, the buffers' closed-loop poles had to be at least ten times higher than the unity-gain frequency of the implemented amplifier. Therefore the AD8000 buffers with a 1.5 GHz bandwidth were used. Figure 6.2 illustrates the designed PCB, and the prepared board with soldered components and chip-on-board preparation of the die.

The test setup used to characterize the performance of the amplifier is shown in Fig. 6.3. This setup was adapted from a Texas Instruments fully differential amplifier THS4521D evaluation module [164]. In Fig. 6.3a a circuit diagram of the amplifier in

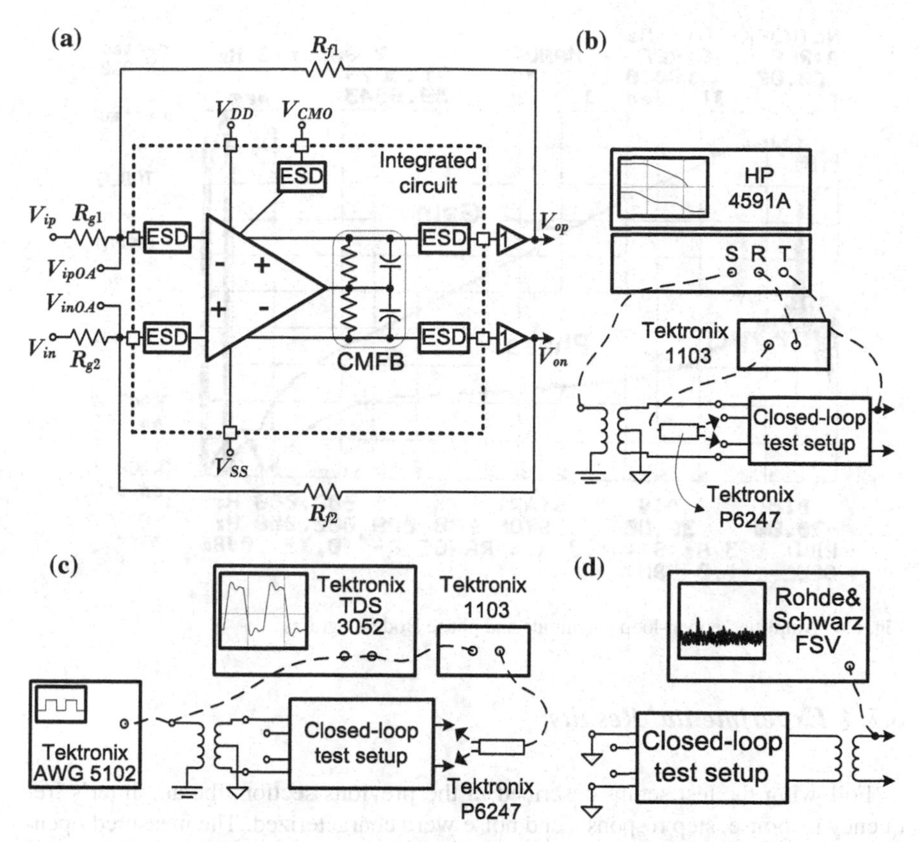

Fig. 6.3 Test setups for the amplifier evaluation: **a** Closed-loop circuit diagram. Setups for **b** open-loop frequency response, **c** step response, and **d** output noise measurement

closed-loop form is shown. For resistors $R_{f1,2}$ and $R_{g1,2}$ a value of 100 Ω was used. Fig. 6.3b shows the setup used to obtain the Bode diagrams (magnitude and phase) of the open-loop gain. The step response of the amplifier was evaluated with the setup of Fig. 6.3c, and finally, Fig. 6.3d shows how the output noise was measured.

The test equipment used to measure the frequency response, the time response, and the noise is given below:

- HP 4195A Network Analyser: frequency response;
- Tektronix AWG510: generate input signals for step response;
- Tektronix TDS3052 Oscilloscope: observe output signals from step response;
- Rohde&Schwarz FSV Signal Analyser: noise measurements;
- Tektronix P6247 Differential Active Probe (and Tektronix 1103 Probe Power Supply): probe differential signals;
- Agilent 6624A DC Power Supply.

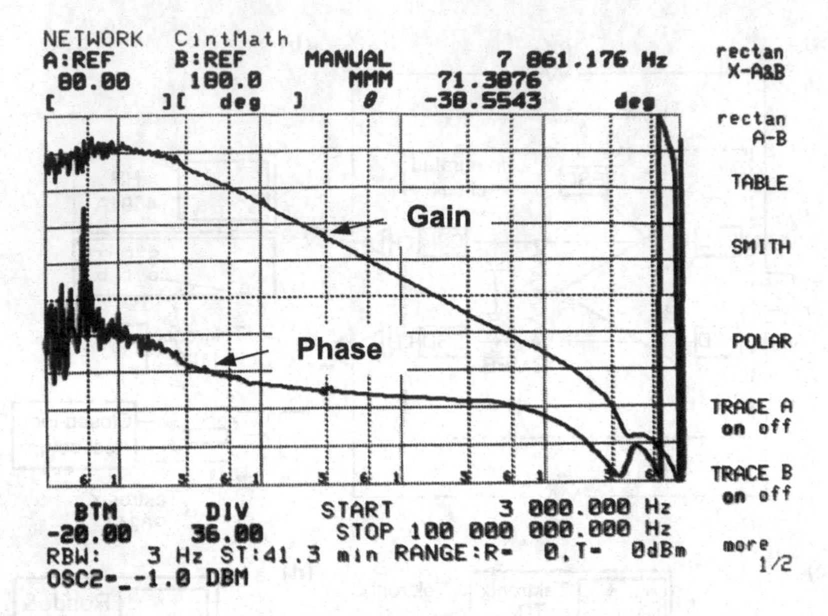

Fig. 6.4 Amplifier's open-loop magnitude and phase Bode diagrams

6.1.4 Experimental Results

Following the test setups described in the previous section, the amplifier's frequency response, step response, and noise were characterized. The measured open-loop magnitude and phase Bode diagrams of the amplifier are shown in Fig. 6.4. For these measurements the amplifier was in unity gain configuration and the active probe was connected to the amplifier's inputs (V_{ipOA} and V_{inOA}). The AC response was then measured between the output of the setup and the amplifier's inputs ($V_{od}/(V_{ipOA} - V_{inOA})$). Due to this setup, the amplifier's inputs were loaded with an extra 200 kΩ || 1.0 pF impedance, while the amplifier's outputs were loaded with the RF pad, the PCB trace (parasitic capacitance much larger than initially anticipated) and the input impedance of the AD8000 ($\approx$ 2 MΩ || 3.6 pF). This extra loading degraded the AC response, especially the PM and the unity gain frequency, which were measured to be less than 45° and 30.4 MHz, respectively. The GBW was extrapolated to be around 35 MHz. Regarding the low frequency (DC) gain of the amplifier, over 71 dB was measured, as indicated in Fig. 6.4. The large gain of the amplifier made it even more difficult to measure, which explains the inaccuracy of Fig. 6.4 at low frequencies, especially in the phase diagram. Regarding extracted layout AC simulations of the amplifier over all 27 PVT corners,[1] the DC gain varied less than 10 % (±5 dB), the unity gain frequency varied less than 55 % (±16 MHz),

[1] Corners considered: *tt*, *ss*, and *ff* for process; 1.2 V±5% for supply voltage, and −40°, 27°, and 85° for temperature.

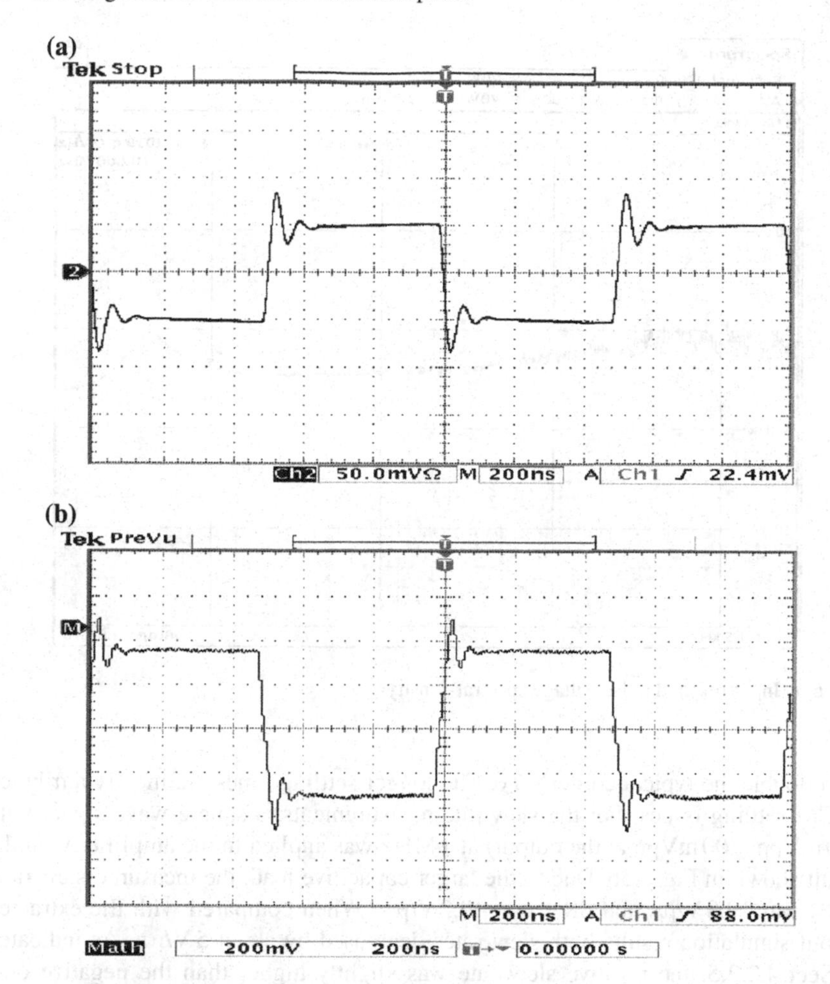

Fig. 6.5 Step responses: **a** Small-signal. **b** Large signal

the current consumption varied approximately 75 % (±75 μA), and the PM less than 40 % (± 15 °). Each parameter was normalized to the result of the typical corner.

To measure the small-signal step response, the loop of the amplifier was closed with a gain of two and a square wave signal with 50 mVpp (100 mVpp at the output) at 1 MHz was applied to the setup's input, with the result shown in Fig. 6.5a. Even with a closed-loop gain of two, the amplifier still denotes some ringing (due to the unexpected larger output capacitance). Measuring the settling time (T_S) proved to be a difficult task given the limited (8 to 9 bits) vertical resolution of the oscilloscope, but at 1 % accuracy, the settling time was measured to be approximately 134 ns, quite far from the (RC extracted) simulated value of 90 ns (for 1 % accuracy). Regarding extracted layout transient simulations in unity-gain configuration (all PVT corners), the T_S was mostly affected by the *ss* corner where, in some cases, it was more

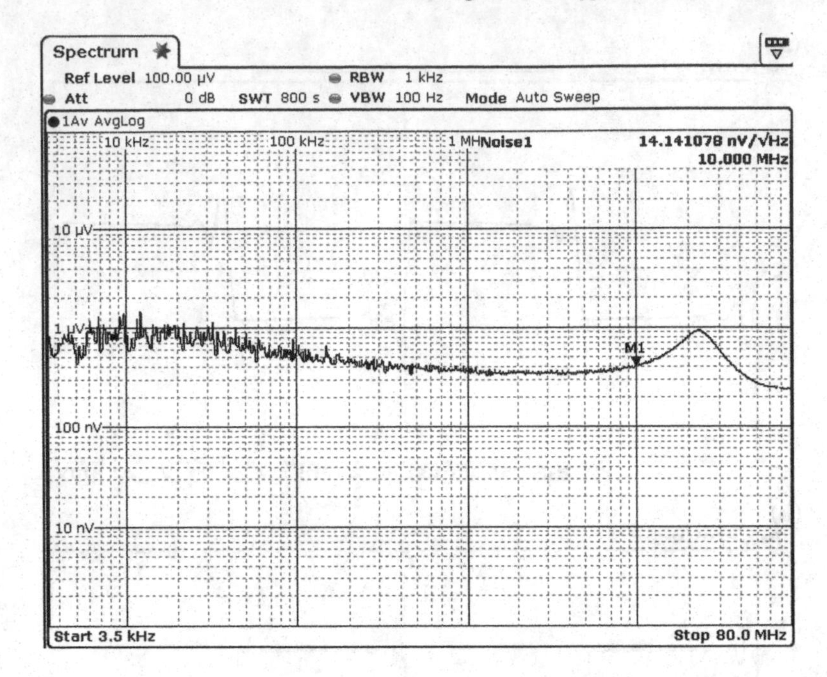

Fig. 6.6 Input-referred noise voltage spectral density

than double the typical corner's T_S. The longer settling times are mostly attributed to CM settling issues. For the slew rate measurements, a square wave signal with 350 mVpp (700 mVpp at the output) at 1 MHz was applied to the amplifier with the result shown in Fig. 6.5b. Due to the larger capacitive load, the measured slew rate, SR^+, was 19.8 V/μs and SR^- was 19.2 V/μs. When compared with the extracted layout simulation results both slew rates decreased by about 5 V/μs. As indicated in Sect. 4.2.2.5, the positive slew rate was slightly higher than the negative one. The noise of the amplifier was characterized in unity gain configuration by shorting the inputs to the input common-mode voltage and a 1:1 differential-to-single-ended transformer was used at the output to measure the differential noise, with the result given in Fig. 6.6. Various averages were taken in order to obtain a more correct reading. For a bandwidth from 3.5 to 80 MHz, the total measured integrated noise was approximately 122 μVrms. The thermal noise level at 10 MHz was 14.14 nV/$\sqrt{\text{Hz}}$ (shown in Fig. 6.6). At frequencies around 25 MHz, there is a visible peaking in the noise characteristic. This is due to the closed-loop peaking, caused by the low PM of the amplifier. A DC blocking filter was used for protection of the signal analyser, so the flicker noise could not be clearly observed. The filtering effect (at low frequencies) of the output transformer is also partly responsible for this. The results of Fig. 6.6 and the noise values presented here are for the complete test setup including 100 Ω R_f and R_g resistors and low-noise buffers used at the amplifier's output (input-referred

Table 6.2 Key experimental summary of the amplifier

Parameter	Value
DC gain	71 dB
GBW[a]	35 MHz
Settling time[a] @ 0.1 V step	134 ns
Settling time accuracy	1 %
Phase Margin[b]	45°
Slew Rate	19.5 V/μs
Input-referred Noise	14.1 nV/$\sqrt{\text{Hz}}$
Power @ 1.2 V	0.110 mW
FoM	1750 MHz · pF/mW

[a] $C_L > 5.5$ pF [b] Assuming a closed-loop gain of two

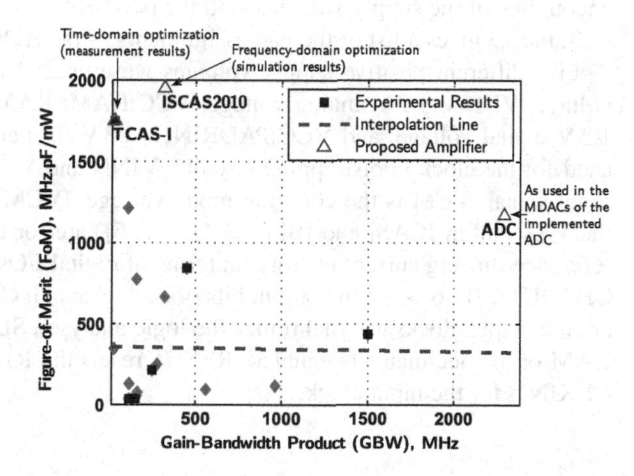

Fig. 6.7 State-of-the-art two-stage opamps with GBW >30 MHz in CMOS technologies. Addition of three different sizings of the proposed amplifier

noise of buffers is 1.6 nV/$\sqrt{\text{Hz}}$). The noise introduced by these circuits play a small role in the total measured noise.

Table 6.2 summarizes the key experimental results obtained for this amplifier.

Given that throughout the work of this book, three amplifiers were implemented: one with measurements of a silicon prototype (the present amplifier, optimized in the time domain) [50], one used in the ADC (described in Chap. 5), and one presented at ISCAS2010 (simulated results only, optimized in the frequency domain) [47], it would be interesting to see their performance compared to the state-of-the-art. Updating Fig. 2.20 by adding the results of these three amplifiers results in Fig. 6.7. It is clearly seen that the performance of the proposed amplifier, even for high GBW (amplifier marked 'ADC'), maintains a FoM within the top amplifiers.

6.2 7-bit 1 GS/s CMOS Two-Way Interleaved Pipeline ADC

6.2.1 Floorplan and Layout

The floorplan of the complete A/D converter is given in Fig. 6.8a. For a better view and understanding, a zoom is made to the floorplan of a stage, shown in Fig. 6.8b. Fig. 6.9 depicts the layout and chip photograph of the complete ADC. Unfortunately, the only visible objects in the chip photograph were the MIM capacitors and the metal-8 connections, therefore, the layout has been overlaid on the photograph for providing a better idea. The core active area of the ADC occupies $840 \times 160\,\mu m^2 = 0.134\,mm^2$. As can be seen in Fig. 6.9b, a large amount of area is dedicated to the decoupling of the supply voltages, and the two SRAM blocks.

Table 6.3 gives a list of the pad assignments of the ADC IC prototype.

Five different positive supply voltages were used: VDDM: 1.2 V mixed-mode voltage, VDDA: 1.2 V analog voltage, VCC:RAM: RAM supply voltage, VDDD: 1.2 V digital voltage, and VCC:PADRING: 3.3 V for padring. VCB is the voltage used for the clock bootstrapping circuits. VINN and VINP are for the differential input signal. VCM is the common-mode voltage. IVCM is the biasing current for the V_{CM} buffer. ICASi and IBi ($i = \{1, \ldots, 5\}$) are for the biasing currents of the reference shifting current mirrors. In terms of digital I/Os, CH1OUT$< 0 : 6 >$ and CH2OUT$< 0 : 6 >$ are the output bits of each channel, CLK1OUT and CLK2OUT are the output clocks to synchronize the logic analyser, SEL determines whether the RAM or the decimator is enabled, RESET resets the RAM controller, and finally, CLKIN is for the input clock.

6.2.2 Layout Considerations

This section will attempt to detail some of the considerations taken during the layout phase of the design.

- The layout was made as symmetric as possible, with two main lines of symmetry: a vertical line between the positive and negative sides of the MDAC's SC network (in each stage) and a horizontal line that runs through the pipeline core. The latter guarantees symmetry between the two interleaved channels. Symmetry of each building block was also guaranteed. The various circuits were laid out as tight as possible thus minimizing parasitic capacitances.
- Separate supply lines and separate pads have been used for the analog, mixed-mode, and digital voltages. The analog voltage lines (V_{DDA} and V_{SSA}) run through the middle of the pipeline, while the mixed-mode and digital lines were laid out as far as possible at the top and bottom of the pipeline (see Fig. 6.8b). To reduce the resistance of each voltage line, two wide metal layers running in parallel (M5 and M7) were used.

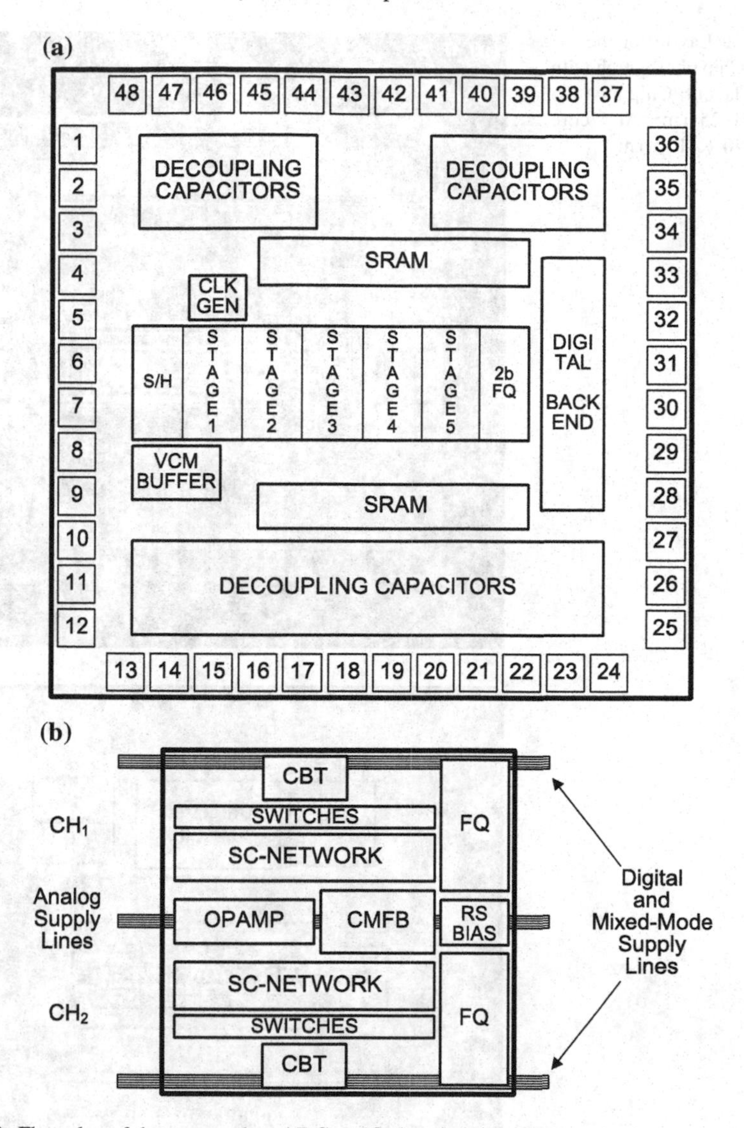

Fig. 6.8 Floorplan of the: **a** complete ADC and **b** time-interleaved stage

- The V_{CB} voltage (used for clock bootstrapping) was laid out near the digital lines, also using two wide metal layers running in parallel (M5 and M7). A dedicated input was used for the V_{CB} voltage.
- The remaining die area was used to place large decoupling capacitors for the supply voltages (40 pF each) and V_{CB} (105 pF).

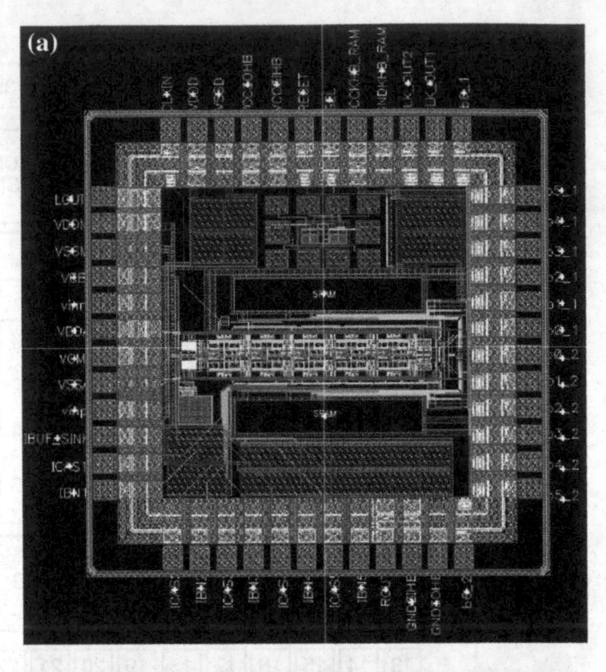

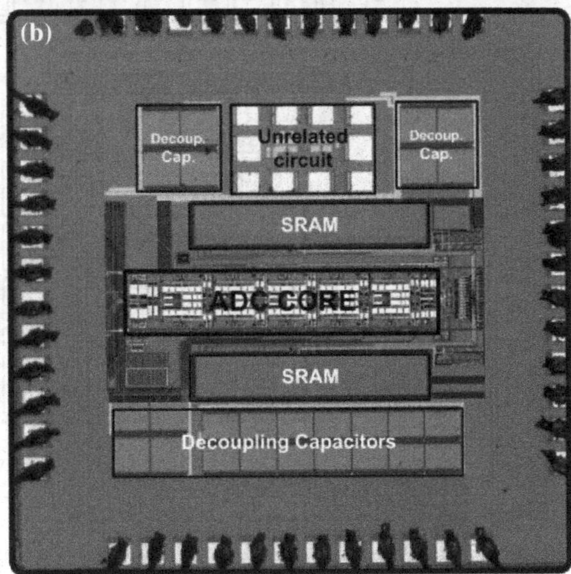

Fig. 6.9 **a** Layout of the ADC. **b** Chip photograph with overlaid layout. Chip area is $1525 \times 1525\,\mu m^2$; ADC core area is $840 \times 160\,\mu m^2$

- All blocks, analog and digital, were shielded by an n-well guard ring. Special attention was devoted to: the amplifiers, the analog part of the flash quantizers, the biasing blocks, and the switches found at the amplifier's inputs.

Table 6.3 Pad assignments of the ADC

1	LCUT	13	ICAS2	25	CH2OUT< 5 >	37	CH1OUT< 6 >
2	VDDM	14	IB2	26	CH2OUT< 4 >	38	CLK1OUT
3	VSSM	15	ICAS3	27	CH2OUT< 3 >	39	CLK2OUT
4	VCB	16	IB3	28	CH2OUT< 2 >	40	GND:RAM
5	VINN	17	ICAS4	29	CH2OUT< 1 >	41	VCC:RAM
6	VDDA	18	IB4	30	CH2OUT< 0 >	42	SEL
7	VCM	19	ICAS5	31	CH1OUT< 0 >	43	RESET
8	VSSA	20	IB5	32	CH1OUT< 1 >	44	VCC:PADRING
9	VINP	21	RCUT	33	CH1OUT< 2 >	45	VCC:PADRING
10	IVCM	22	GND:PADRING	34	CH1OUT< 3 >	46	VSSD
11	ICAS1	23	GND:PADRING	35	CH1OUT< 4 >	47	VDDD
12	IB1	24	CH2OUT< 6 >	36	CH1OUT< 5 >	48	CLKIN

- To minimize the timing mismatch, in the S/H all the lengths and couplings of the clock/phase lines were equalized. The same was performed for the positive and negative lines of the MDAC that connect to the amplifier's input switches. The main clock phase lines, from the nonoverlapping clock generator, were also matched.
- To adjust the gain of the MDAC, 3 fF capacitors were placed at the MDAC's critical nodes (see $C_{p21,22}$ of Fig. 3.5).
- Dummy block layers were added to the design over all critical circuits, namely: amplifiers, MDACs' SC network, flash quantizers' sampling capacitors, and CMFB circuits. This was performed in order to avoid the foundry adding layers (due to metal coverage rules) near or over critical circuits of the converter.
- Regarding the digital backend, the synchronization logic was placed along the pipeline, to avoid the accumulation of parasitic capacitances on long lines. The last bank of flip-flops of this block (the time-alignment flip-flops) was placed at the end of the pipeline. The time-interleaved decimator was placed after the final 2-bit stage in the middle of the pipeline. The correction logic and output buffers were placed in the middle of the pipeline after the decimator.

6.2.3 PCB and Test Setup

To evaluate the performance of the implemented ADC, a PCB was designed using EAGLE layout editor. It was designed with four copper layers: top layer for routing, second layer was a ground plane, third layer was used for the supply voltages, and the bottom layer had routing and was used as another ground plane. On-board components consisted of a voltage reference (ADR440), three low noise opamps for driving the CM voltages ($V_{CMinput}$ for the input signal and V_{CM} for the chip) and V_{CB} (LMP7732), one 1:1 wideband transformer for single-to-differential conversion (ADT1-1WT), and a digital buffer (SN74ALVCH16244DL). Four supply voltages were used on-board, analog 1.2 V for the analog and mixed-mode voltages

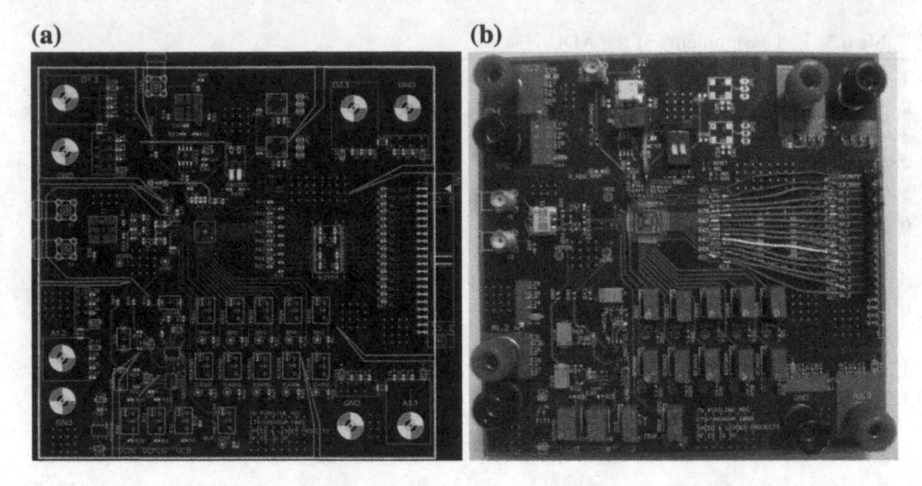

Fig. 6.10 Designed PCB: **a** Layout. **b** Prepared board with chip-on-board soldering of the die

of the chip, digital 1.2 V for the digital and RAM supplies of the chip, analog 3.3 V for components ADR440 and LMP7732, and digital 3.3 V for the digital buffer. All supply voltages were decoupled globally using capacitors with various capacitance values (10 nF – 10μF), an inductor, and again locally (near the chip) with more capacitors (of the same values). The CM voltages and V_{CB} were also decoupled with large capacitors (330 μF). All currents were generated and controlled with variable resistors. The RESET and SEL signals were controlled with a DIP switch. The PCB was prepared for chip-on-board soldering of the die.

Figure 6.10 illustrates the designed PCB, and a prepared board with soldered components and chip-on-board soldering of the die. As can be seen from Fig. 6.10b, not all the above mentioned components were used. The digital buffer was eventually bypassed because it caused instability in the PCB's supply voltages (there was some difficulty soldering this component which could have caused shorted paths or damaging of the chip). Bypassing this component was not a problem because the rate of the output bits was low (due to decimation) and the on-chip digital pads had enough strength to drive the logic analyser's measuring pods. A simplified diagram of the setup employed to test and measure the performance of the ADC is shown in Fig. 6.11. A list of the equipment used during the testing process is given next.

- Agilent 6624A DC Power Supply: supplied the four voltages, 1.2 V (analog and digital) and 3.3 V (analog and digital);
- Marconi Instruments 2041 and HP 8753ES: generated the input signal;
- Stanford Research Systems CG635: generated the clock signal;
- Agilent 16702B Logic Analyser: captured the digital data;
- Agilent 34401A Multimeter: voltage verification and current measurement;
- Tektronix TDS3052 Oscilloscope: verification of node voltages, input and clock signals;
- Allen Avionics Bandpass Filters: used to filter the input signal;

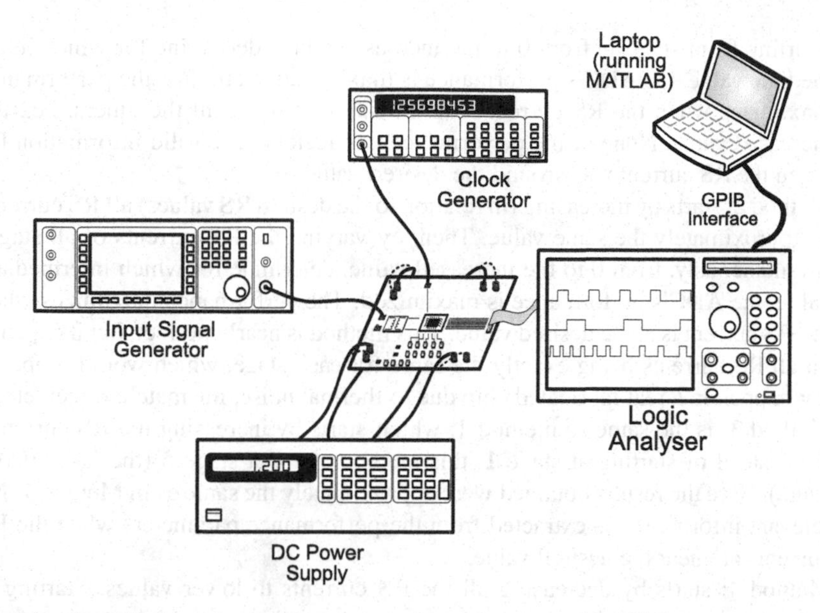

Fig. 6.11 Simplified test setup used to measure the performance of the ADC

- Laptop running MATLAB: get data from logic analyser and compute FFT, DNL/INL, etc.

6.2.4 Employed Methodology for Tuning the Reference Currents

It was not an easy task to define a valid methodology to tune the currents in order to have the correct reference shifting (RS), which in turn would maximize the performance of the ADC. For this task, an ideal pipeline converter designed in MATLAB was used to arrive at an approximate method for current tuning. Remember that the currents for the current mirrors of each stage were generated and controlled off-chip, given the proof-of-concept nature of the implemented ADC.[2]

The ideal converter, used in MATLAB, was ideal in the sense that no building block had deviations from its nominal operation, except for the RS currents, which were under test. The ADC parameters used as a means to determine the converter's performance were DNL, INL, SFDR, SNR, THD, and ENOB. The following gives a brief description of the tested methods:

- Method 1: starts by increasing (in relation to the desired RS values) all RS currents to approximately the same value. Then, by only varying one RS current at a time

[2] Naturally, in a redesign of this ADC, the generation of all required reference currents will be provided and automatically adjusted on-chip. As mentioned before, this can be achieved through a SC reference current generator [165] or using a replica MDAC together with a servo-loop (see Appendix A). Note that, no self-calibration will be required.

(starting from stage 1) from 0 to the increased value, determine for which intermediate value, the ADC's performance is maximized. Naturally, the performance maximizes when the RS current is at the same value of all the others, i.e., the increased value. None of the performance parameters gave valid information for when the RS current was around the desired value.

- Method 2: starts by increasing (in relation to the desired RS values) all RS currents to approximately the same value. Then, by varying the RS currents of all stages simultaneously, from 0 to the increased value, determine for which intermediate value, the ADC's performance is maximized. The performance maximizes when the RS current is at the desired value. This method is nearly perfect, but it depends on all RS currents being exactly the same for each stage, which would probably not happen in a real fabricated chip due to thermal noise, mismatch, offset, etc.
- Method 3: is the same as method 1, which starts by increasing the RS currents, but instead of starting at stage 1, this method starts at stage 5 (the last MDAC stage). Here the results obtained were approximately the same as in Method 1. No relevant information is extracted from the performance parameters when the RS currents are near the desired value.
- Method 4: starts by decreasing all the RS currents to lower values. Starting at stage 1 (maintaining all other currents constant) and by varying its current from 0 to a value way passed the desired one, the DNL minimizes and the THD and SFDR maximize when the RS current is at (or near) its desired value. ENOB maximizes just below the desired value because of the SNR. Maximization of the THD and SFDR occur for the desired value because, for low RS values, there is a large amount of distortion due to residue saturation throughout the pipeline, and by adjusting the RS current of the MSB stage, valid information can be extracted. Step 2 would be to leave the RS current of stage 1 at the value that maximizes THD and SFDR (and minimizes DNL), and vary the RS current of stage 2, and so on.
- Method 5: starts by decreasing all the RS currents to lower values. Starting at stage 5 (maintaining all other currents constant) and by varying its current from 0 to a value way passed the desired one, no valid information was extracted. This is due to the incapability of the least significant stages to produce a substantial (visible) effect on the ADC's performance.

During the testing stage of the design, Method 4 was adopted. Testing initiated by adjusting the input signal (frequency, amplitude and its common-mode value), the clock frequency, V_{CM} voltage, 1.2 V supply voltages, and the currents for the current mirrors that generate the cascode voltages (V_{CAS}) of all stages. Then Method 4 was put in action by tuning the currents for the current mirrors that generate the biasing voltages (V_B), that ultimately determine the amount of reference shifting. Various iterations of Method 4 were done, i.e., start at stage 1, go on to the next stage until stage 5, and then return to stage 1 for finer tuning, and so on. This process was repeated until the ENOB was maximized, which was the final objective.

6.2.5 Experimental Results

The proposed ADC was designed to operate at a sampling frequency of 1 GS/s. This rate was only accomplished at the schematic level, where an ENOB of 6.2 bits was achieved (results shown in Sect. 5.3.10). RC extracted layout simulations showed that the MDAC's output characteristic had a nonlinearity when switching from X (or Z) mode to Y mode and vice-versa. The nonlinearity has mostly to do with the impedance of the current source, which was not high enough to suppress input signal modulation at the current source's output node. Besides this, even harmonics were higher, which could have resulted from mismatch between the current sources (amplifier's CMFB circuit was unable to adjust the differential outputs) and/or different current reference shifting for X and Z modes. This distortion limited the ENOB to about 4.6 bits at 1 GS/s. Reducing the sampling frequency to 700 MS/s and by solving layout related and current source issues, it was possible to increase the ENOB to 5.4 bits. Due to the tape-out deadline, no further corrections and improvements were possible. Therefore, the best result going into the testing stage of the design, was an ENOB of 5.4 bits at 700 MS/s (with a low frequency input).

After some initial testing of the silicon prototypes, it was clear that the converter had a duty-cycle error, caused by the lack of an on-chip duty-cycle restorer.[3] This fact in addition to the other mentioned limitations, reduced the maximum achievable sampling frequency to 640 MS/s, achieving a similar ENOB to that of the last RC extracted simulation. Therefore all the measured results that follow are for $F_S =$ 640 MS/s.

Figure 6.12 illustrates the DNL and INL of the time-interleaved ADC at $F_S =$ 640 MS/s. The results show that a good linearity is achieved with a maximum DNL of $+0.57/-0.52$ and a maximum INL of $+1.12/-0.97$. The static linearity of the ADC was measured using a sine wave code density test with 2^{18} samples [42].

To measure the dynamic linearity of the converter, a low and high frequency input signal was used. Figure 6.13a depicts an 8192-point FFT for an input frequency of 10.7 MHz (this particular value was used due to the bandpass filter employed) and $A_{in} = -0.5$ dBFS . Due to decimation by a factor of 15, the frequency axis is limited to $(F_S/2)/15 \approx 21.3$ MHz. As should be noticed, the noise floor's range is quite small, this is because 32 averages were taken for all FFTs. The plot of the figure indicates the results obtained, as well as, the location of the 2nd to the 6th harmonics and the time-interleaved (TI) spur. The THD was measured accounting thirteen harmonics, but only six are shown to avoid loading the plot. The TI spur, which appears just left of the fundamental, arises due to mismatches between the gain and timing of the two interleaved channels. At this input frequency, the spur is low with a power of −57 dB.

[3] Although the clock, provided externally by the CG635 clock generator, may have a precise duty-cycle, the clock input pad (Schmitt-triggered) has unmatched rising and falling times.

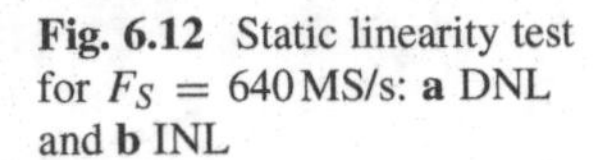

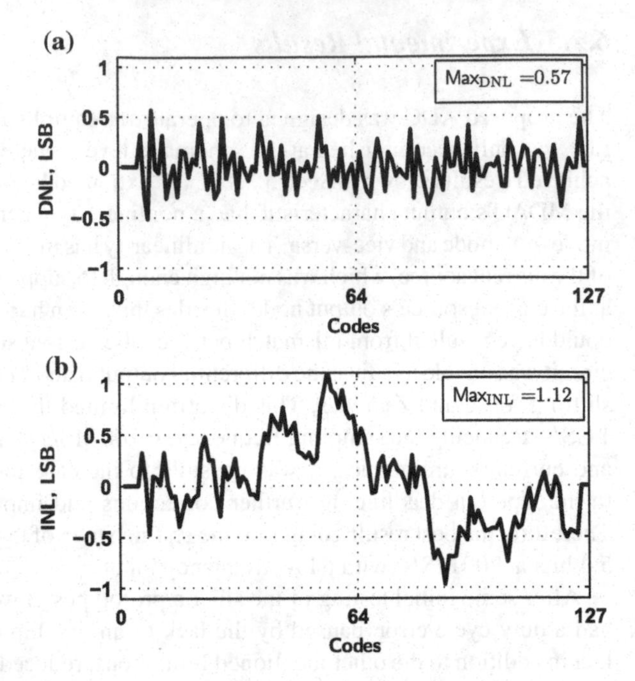

Fig. 6.12 Static linearity test for F_S = 640 MS/s: **a** DNL and **b** INL

For the high frequency test, an $F_S/2 \approx 319\,\text{MHz}$, A_{in} =-0.5 dBFS input was used, with the results depicted in Fig. 6.13b. Again, an 8192-point FFT averaged 32 times was used. Comparing the results for the low frequency input, the THD degraded 1.1 dB, the SNR degraded 1.5 dB (mostly caused by the rise of the TI spur), the SFDR decreased 5.7 dB (also limited by the TI spur), which resulted in a final SNDR of 32.2 dB and an ENOB of 5.1 bits. The lack of a duty cycle restorer is well present in this FFT, as the timing error caused the TI spur to rise, degrading both the SNR and SFDR. Regarding jitter, by removing the energy of the TI spur from both FFTs and recalculating the SNR, an equal SNR is obtained, meaning that the noise floor did not increase (which would have been caused by jitter). Therefore, the degradation in SNR from a low frequency to a high frequency input is mainly caused by the TI spur.

The total power consumed by the ADC (including reference circuitry) was 15.5 mW which resulted in a FoM of 706 fJ/conv.-step. This power excludes the V_{CM} buffer (not optimized for 640 MS/s), the digital output buffers, the decimator, the RAM, and its controller.

Figure 6.14a depicts the SNDR and SFDR for various input signal frequencies for F_S = 640 MS/s, where a linear degradation is noticeable. Fig. 6.14b depicts the SNDR and SFDR for various sampling frequencies and f_{in} = 1 MHz. The results are approximately constant for F_S < 640 MS/s. The ADC maintains a SFDR above 6 bits for input signals up to $F_S/2$, and above 7 bits for sampling frequencies up to 700 MS/s

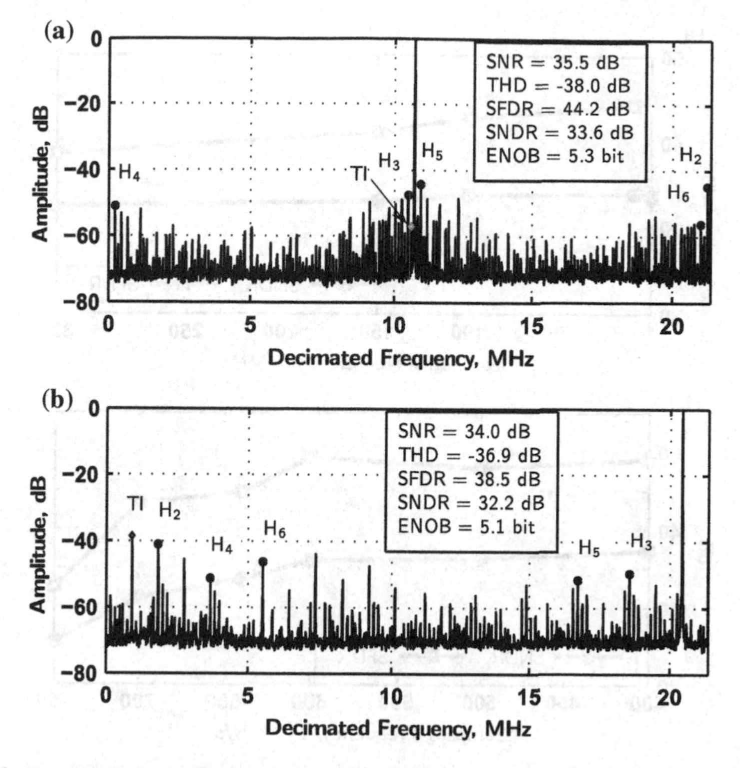

Fig. 6.13 Dynamic linearity test for $F_S = 640$ MS/s and $A_{in} = -0.5$ dBFS. 8192-point FFT for: **a** 10.7 MHz input and **b** $F_S/2$ input

(for low frequency inputs). Regarding ENOB, the ADC maintains approximately 5 bits for sampling frequencies up to 700 MS/s (for low frequency inputs).

Figure 6.15 depicts the total power consumption of the converter and the reference shifting current per stage for various sampling frequencies. Each figure has an interpolation line and its inclination is indicated, showing how the techniques proposed in this book scale with the sampling frequency. For F_S=640 MS/s the power consumed was 15.5 mW with 57.8 μA per stage used for reference shifting.

The distribution of the power consumption of the overall converter is illustrated in Fig. 6.16, using simulation results obtained for $F_S = 640$ MS/s. The pie chart shows that, as expected, the opamps are still the most power consuming blocks of the converter (notice that the pipeline ADC has not been scaled down), followed by the flash quantizers and the clock bootstrapping circuits. Measurement results show that 90 % (13.9 mW) of the power is attributed to analog and mixed-mode blocks, while the remaining 10 % (1.6 mW) is used to power digital circuitry. To conclude this section, Table 6.4 summarizes the key experimental performance characteristics of the ADC and a comparison with the state-of-the-art is carried out in Fig. 6.17 by updating the original data of Fig.2.21. Figure 6.17 shows results from CMOS

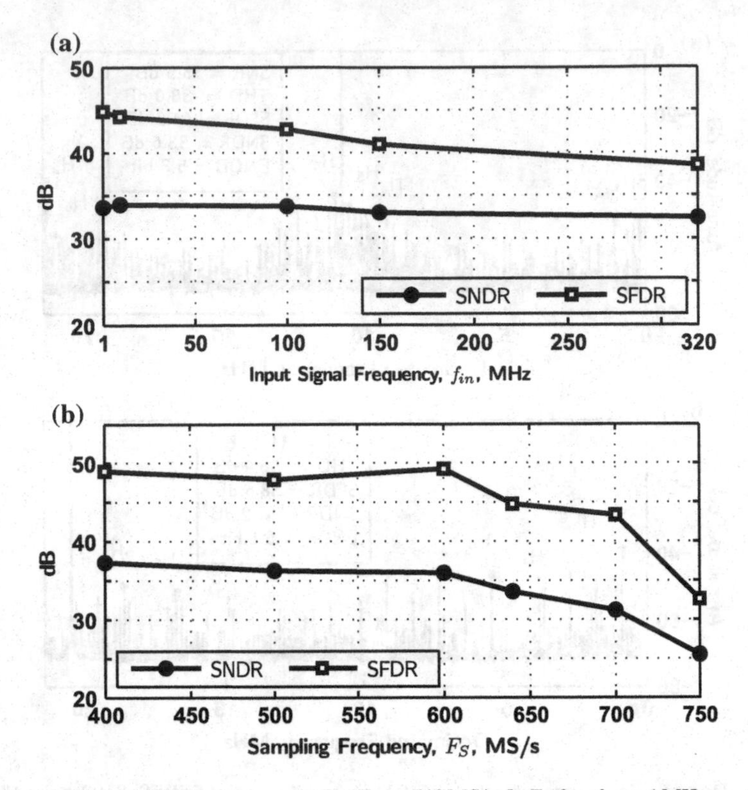

Fig. 6.14 SNDR and SFDR versus: **a** f_{in} for $F_S = 640\,\text{MS/s}$. **b** F_S for $f_{in} = 1\,\text{MHz}$

ADCs with resolutions of 6–8 bits, with ENOB > 5 bits, and sampling frequencies >200 MS/s. As can be seen in Fig. 6.17a, the proposed ADC is situated better (in terms of FoM) than the average, and ranks fourth in this state-of-the-art. The analysis of the state-of-the-art must be carried out with some caution, which is well illustrated by the interpolation line, i.e., as the sampling frequency increases, it becomes difficult to push the FoM. An interesting example of this would be to only consider ADCs with F_S >500 MS/s, and in this case, the proposed ADC would rank second. In terms of accuracy (Fig. 6.17b), the proposed ADC drops 1.9 bits for a Nyquist input, mainly due to the timing mismatch, which becomes the dominant limitation in terms of dynamic range. Therefore, the performance of the ADC is largely penalized due to the lack of a duty-cycle restorer.

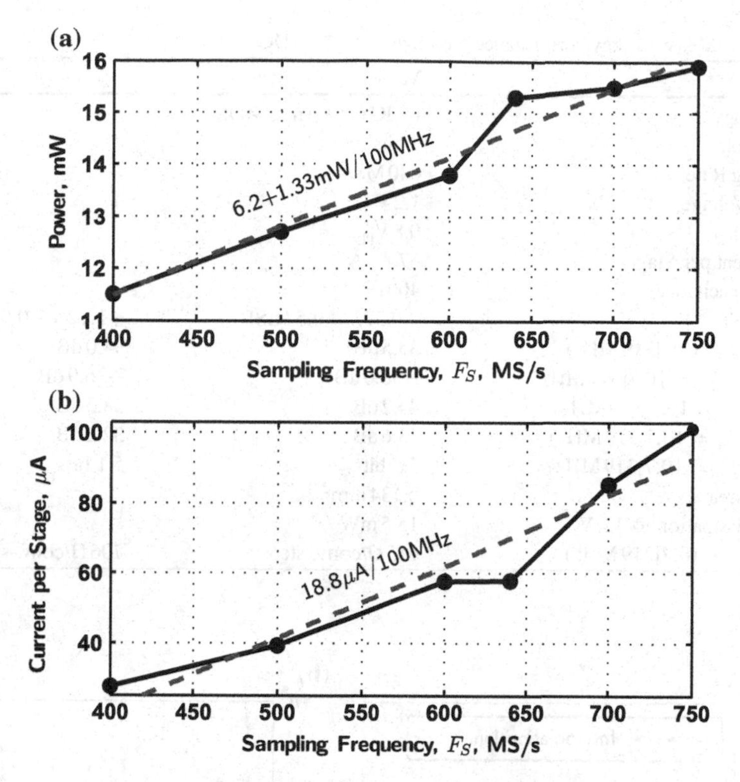

Fig. 6.15 **a** Total power consumption and **b** reference shifting current per stage scalability for various sampling frequencies

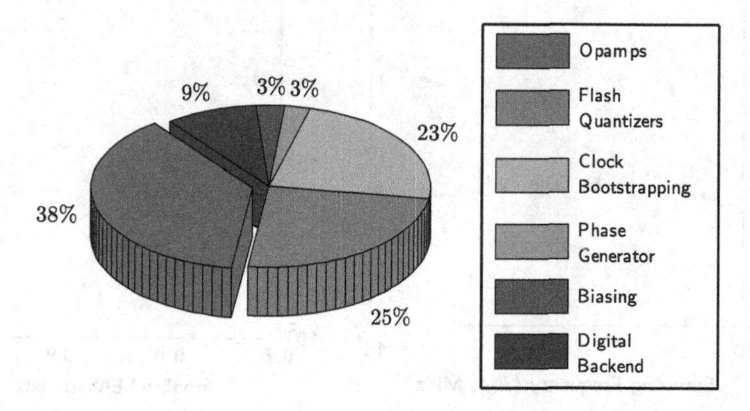

Fig. 6.16 Power consumption distribution of the overall time-interleaved ADC

Table 6.4 Measured key performance parameters of the ADC

Parameter	Value	
Technology	UMC 0.13 μm CMOS	
Resolution	7 bits	
Sampling Rate, F_S	640 MS/s	
Supply Voltage	1.2 V	
Input Range	0.8 $V_{pp-diff.}$	
RS Current per Stage	57.8 μA	
Input Capacitance	400 fF	
DNL | INL	+0.57/ − 0.52 LSB	+1.12/ − 0.97 LSB
SNR (f_{in} = 10.7|319 MHz)	35.5 dB	34.0 dB
THD_{13H} (f_{in} = 10.7|319 MHz)	−38.0 dB	−36.9 dB
SFDR (f_{in} = 10.7|319 MHz)	44.2 dB	38.5 dB
SNDR (f_{in} = 10.7|319 MHz)	33.6 dB	32.2 dB
ENOB (f_{in} = 10.7|319 MHz)	5.3 bits	5.1 bits
Active Area	0.134 mm^2	
Power dissipation @ 1.2 V	15.5 mW	
FoM (f_{in} = 10.7|319 MHz)	615 fJ/conv.-step	706 fJ/conv.-step

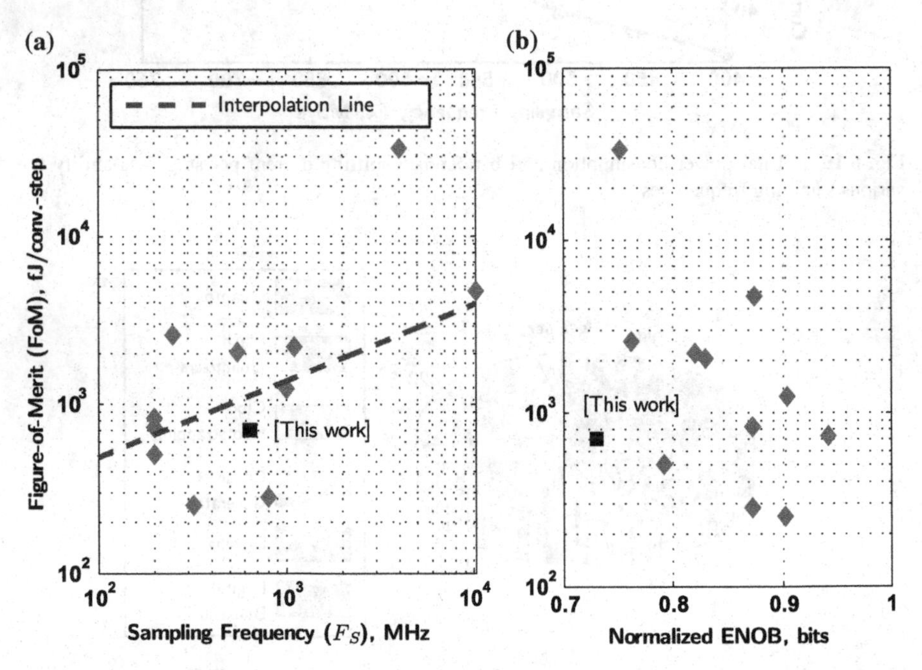

Fig. 6.17 The state-of-the-art of 6–8 bit MDAC-based ADCs with ENOB >5 bits and F_S > 200 MS/s in CMOS technologies: **a** FoM versus F_S. **b** FoM versus normalized ENOB

Chapter 7
Conclusions

7.1 Final Remarks

This book has explored various design techniques with the purpose of enhancing the power and area efficiency of building blocks mainly to be used in MDAC-based ADCs. The developed techniques mostly rely on improving the circuits in the analog domain, while not using any type of digital or mixed-signal calibration. It was out of the scope of this book to propose any type of digital assistance schemes or algorithms, but rather develop novel techniques that improve the technology's analog capabilities.

To meet this goal a number of analog techniques were developed or used to enhance the performance of the various circuits proposed in this work. The following highlights and summarizes the most important:

- self-biasing for improved PVT insensitivity;
- inverter-based design for improved power/speed ratio, i.e., better efficiency;
- unity feedback factor MDACs for improved efficiency;
- capacitor mismatch insensitive MDACs for improved matching;
- current-mode reference shifting for MDACs, which eliminate reference voltage circuitry (including associated buffers and decoupling capacitors), therefore improving power and area efficiency of the overall ADC.

The combination of some of these techniques allowed designing four circuits, three of which were integrated in a larger block, demonstrating their effectiveness, functionality, and performance. These circuits are:

- a 1.5-bit flash quantizer employing self-biasing and inverter-based design;
- an operational transconductance amplifier (OTA) with self-biasing and inverter-based design;
- an MDAC with unity feedback factor, insensitive to capacitor mismatch, and current-mode reference shifting capability;
- a multiply-by-two-amplifier (MBTA) with unity feedback factor and insensitive to capacitor mismatch.

M. Figueiredo et al., *Reference-Free CMOS Pipeline Analog-to-Digital Converters*, Analog Circuits and Signal Processing, DOI: 10.1007/978-1-4614-3467-2_7,

The first three itemized circuits were integrated in a high-speed medium-low resolution pipeline ADC, designed and implemented in a standard 0.13 μm CMOS technology (without using any special devices or options) with experimental results given in Sect.6.2. This IC prototype allowed assessing the functionality and performance of these circuits. An IC prototype of the amplifier (exclusively) was also designed and implemented in the same technology, and tested, with experimental results given in Sect.6.1.

For each of the mentioned circuits a complete description and various theoretical analyses are carried out. A comparison with their respective conventional counterparts and a final comparison with the state-of-the-art is also performed. Moreover, advantages and limitations of each circuit is described, and for most of the limitations, viable solutions are given. Finally, simulation results, and in the case of the amplifier and pipeline ADC, experimental results are presented.

Looking back at the research goals and objectives of this work outlined in the introductory chapter of this book, practically all goals were accomplished except for the fact that the performance of the ADC did not push the state-of-the-art as was initially planned and desired. It was probably too audacious to integrate three novel circuits into a pipeline ADC, and still expect to push the limit in ADC performance. Nevertheless, interesting results have been achieved, that prove the practicality, functionality and performance of each of the designed circuits.

In the continuation of this work further research can be carried out in various aspects of the designed circuits. In particular, the following ideas seem very promising to pursue.

- Redesign of the pipeline ADC, pushing the 1 GS/s (and beyond) limit, solving all issues: MDAC's input-output characteristic nonlinearity and inclusion of a duty-cycle restorer (or a more sophisticated clocking scheme). Integration of a circuit that generates the reference shifting currents (e.g., SC current reference generator or replica MDAC with a servo-loop). Reduce thermal noise of current sources (for current-mode reference shifting) by adding decoupling capacitors. Remove the S/H block and scale-down the pipelined-stages to reduce power (for instance, use one opamp sizing for the first two MDACs and a down-scaled version for the remaining MDACs). Optimize power of clock bootstrapping circuits and flash quantizers, as these occupy too much of the total power consumption.
- Target higher resolution (10–12 bit) ADCs, to benefit from the advantages of the proposed MBTA and MDAC circuits being insensitive to capacitor mismatch. Simultaneously, the current-mode reference shifting (in the MDAC case) capability at these resolutions can also be assessed.
- Analyse the portability of the proposed techniques, especially the MDAC, to deeper sub-micron CMOS nodes, such as 65 and 40 nm.
- Develop a new technique to self-bias both stages of the amplifier, improving CM speed and stability at the output. This will also lead to a faster differential settling, possibly improving the amplifier's efficiency.
- Derive a time domain figure-of-merit (FoM) for multi-stage amplifiers that takes into consideration the settling time, the settling accuracy, and the input step

amplitude. This will allow for a better comparison of class-A and class-AB amplifiers, as well as, single and multi-stage amplifier architectures.

- Extend the self-biasing techniques to other circuits, such as, clocking circuits, digital circuitry, ring oscillators, VCOs, PLLs, RF circuits, among others. Some of this work has partially been done, such as, clock generators [46], multi-stage amplifiers with common-gate devices [147], LNAs [38], and ring oscillators [154].

Appendix
Solution for Current Reference Shifting Integration

The following solution solves two issues of the implemented ADC described in Chap. 5: integration of the currents for reference shifting and reduction (and control) of the reference shifting period to allow linear settling before the end of the amplification phase [127]. As already described in Sect. 3.2, a current-mode reference shifting technique is employed to substitute the voltage reference circuits. The objective is to emulate the effect of a reference voltage (V_{REF}), which can be achieved using a capacitor, a current, and a certain amount of time, i.e., $\Delta V = I\Delta t/C$. In an SC-MDAC circuit, two of the three mentioned quantities are readily available: the capacitor is the one found in the opamp's feedback loop, the maximum time available is the amplification phase, and only a current source needs to be added to the circuit (substituting the reference voltage). To determine the correct amount of current for reference shifting, a global reference shifting controller, composed of a replica of a pipeline stage and a discrete SC-integrator, is used (Fig. 1.1). These two circuits operate together to generate the biasing voltages for the cascode current mirrors and sources of all the ADC's stages. Note that, the V_{REF} of Fig 1.1 does not need to be buffered.

A.1 Replica Stage

As the name indicates the replica stage is an exact copy of a pipelined stage. Any deviations or changes to the circuit's nominal operation will be reflected in this replica stage, and therefore, corrected. The replica circuit helps minimizing issues caused by charge injection of the MDAC's current control switches, duty-cycle deviation between the main phases, and undesired parasitic effects (opamp's input, etc.). Scaling of the replica stage is possible, with respective degradation of reference shifting accuracy. Given that the ADC's performance depends on the matching between the replica and the ADC's stages, caution must be taken during transistor sizing to avoid large mismatches.

M. Figueiredo et al., *Reference-Free CMOS Pipeline Analog-to-Digital Converters*, Analog Circuits and Signal Processing, DOI: 10.1007/978-1-4614-3467-2,

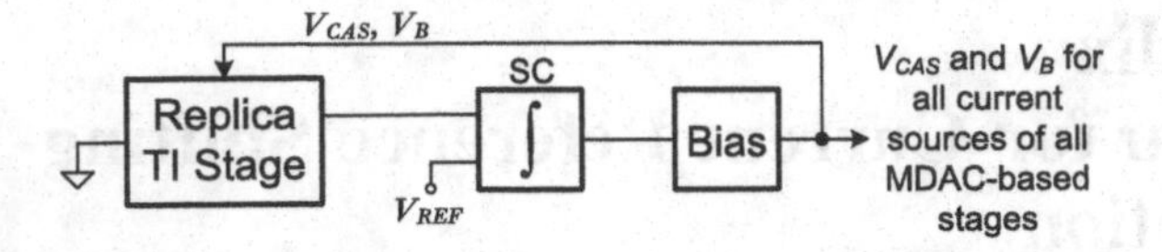

Fig. 1.1 Block diagram of the global reference shifting current controller

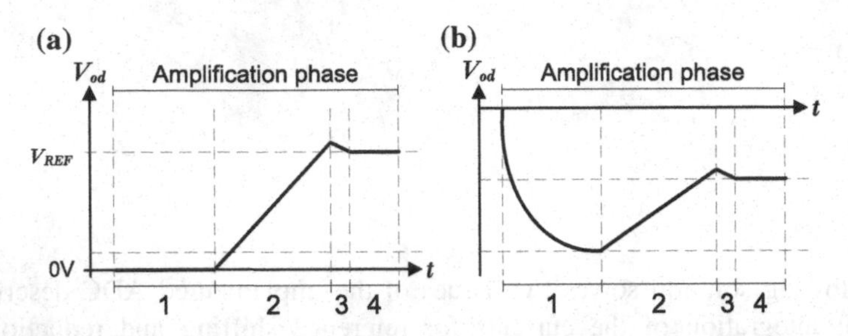

Fig. 1.2 Timing diagrams: **a** MDAC of the replica stage. **b** MDAC of a pipelined stage. 1: Gain of two starts, 2: Current reference shifting, 3: Charge injection, and 4: Settling

The MDAC circuit of the replica stage has its inputs tied to analog ground, therefore, it always (differentially) samples zero. The flash quantizer has its inputs tied to a differential voltage that forces X always to be high, i.e., it always forces the MDAC to perform reference shifting (summing the reference). The outputs of the replica stage are connected to the integrator and compared with the desired reference shifting value, which depends on the full-scale range of the ADC. The timing diagram of Fig. 1.2a illustrates the output of the replica stage, which has a gain of two of 0 V (because of the 0 V input voltage), reference shifting ($X = 1$) and a final period to settle. Figure1.2b shows a possible output voltage waveform of a pipelined stage, with a gain of two of the sampled voltage, followed by reference shifting and a final settling period.

A.2 Current Reference Shifting Period Controller

The controller, responsible for generating a shorter phase (shown in Fig.1.3), initiates operation with capacitor, C, discharged. The input, represented by X, Y, or Z, goes from 0 to 1, which causes the output (XYZ^*) to rise, because INV_2's output is also high. This output connects to the MDAC's transmission gates and current sinking is initiated. Simultaneously, M_1 turns on and charges C until INV_2's output becomes zero. This causes XYZ^* to fall, terminating the current sinking period. When the input falls, at the end of ϕ_2, M_2 turns on, discharging the capacitor.

The capacitor and $M_{1,2}$'s ON-resistance (R) determine the charging/discharging time constant, $\tau = RC < 1/2F_S$, where F_S is the ADC's sampling frequency.

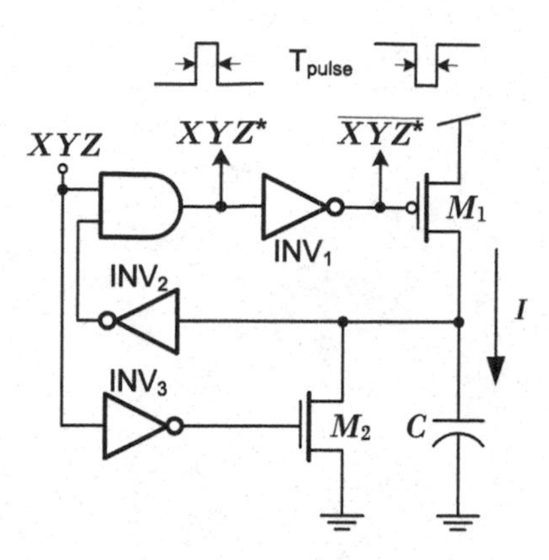

Fig. 1.3 Circuit diagram of the current reference shifting period controller

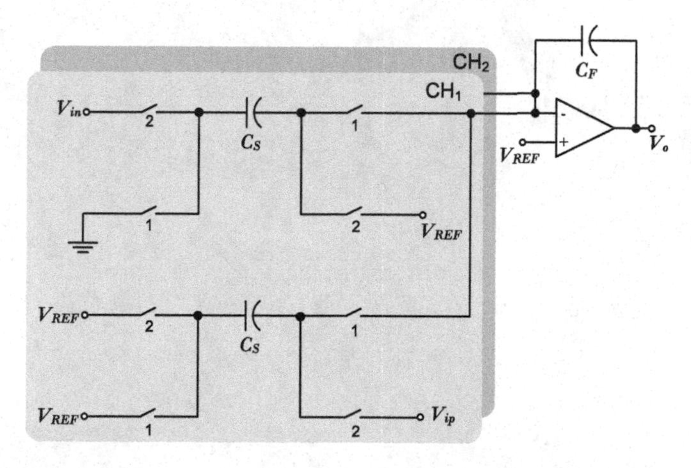

Fig. 1.4 Integrator circuit for time-interleaved stages. For single channel ADCs, one of the channels of the integrator may be discarded

A.3 Discrete SC-Integrator

The SC-integrator, as shown in Fig.1.4, is based on the conventional parasitic-insensitive architecture. The circuit diagram illustrates an integrator for a time-interleaved ADC with two channels. For single channel ADCs, one of the channels of the integrator is removed. The outputs of the replica stage are connected to the integrator and are compared (differentially) with the desired reference shifting voltage. The output of the integrator drives a current mirror of the biasing circuit to generate two biasing voltages, V_{CAS} and V_B. These voltages drive the current sources of the replica stage (and the current sources of all stages of the ADC) in a negative feedback loop. This analog calibration scheme allows to continuously adjust the biasing voltages until the desired reference shifting is achieved, taking into account all the nonidealities of a pipeline stage.

References

1. Abo AM (1999) Design for reliability of low-voltage, switched-capacitor circuits. PhD thesis, University of California at Berkeley
2. Achigui H, Fayomi C, Sawan M (2003) A DTMOS-based 1 V opamp. In: Proceedings IEEE International Conference Electronics, Circuits, Systems (ICECS), pp 252–255
3. AD (2011) Analog Devices AD9484. http://www.analog.com/en/analog-to-digital-converters/ad-converters/ad9484/products/product.html
4. Ahmed I, Johns D (2007) An 11-bit 45MS/s pipelined ADC with rapid calibration of DAC errors in a multi-bit pipeline stage. In: Proceedings of European Solid-State Circuits Conference (ESSCIRC), pp 147–150
5. Ahuja B (1983) An improved frequency compensation technique for CMOS operational amplifiers. IEEE J Solid-State Circuits 18(6):629-633
6. Alpman E, Lakdawala H, Carley L, Soumyanath K (2009) A 1.1V 50mW 2.5GS/s 7b time-interleaved C-2C SAR ADC in 45nm LP digital CMOS. In: IEEE International Solid-State Circuits Conference (ISSCC) Digest Technical Papers
7. Assaad R, Silva-Martinez J (2009) The recycling folded cascode A general enhancement of the folded cascode amplifier. IEEE J Solid-State Circuits 44(9):2535–2542
8. Bajdechi O, Huijsing J (2002) A 1.8-V ΔΣ modulator interface for an electret microphone with on-chip reference. IEEE J Solid-State Circuits 37(3):279–285
9. Baker RJ (2010) CMOS : circuit design, layout, and simulation, 3rd edn. Wiley-IEEE Press, Hoboken
10. Bazes M (1991) Two novel fully complementary self-biased CMOS differential amplifiers. IEEE J Solid-State Circuits 26(2):165–168
11. Black WC, Hodges DA (1980) Time interleaved converter arrays. IEEE J Solid-State Circuits 15(6):1022–1029
12. Borghetti F, Nielsen JH, Ferragina V, Malcovati P, Andreani P, Baschirotto A (2006) A programmable 10b up-to-6MS/s SAR-ADC featuring constant-FoM with on-chip reference voltage buffers. In: Proceedings of European Solid-State Circuits Conference (ESSCIRC), pp 500–503
13. Boulemnakher M, Andre E, Roux J, Paillardet F (2008) A 1.2V 4.5mW 10b 100MS/s pipeline ADC in a 65nm CMOS. In: IEEE International Solid-State Circuits Conference (ISSCC) Digest Technical Papers, pp 250–611
14. Brewer R, Gorbold J, Hurrell P, Lyden C, Maurino R, Vickery M (2005) A 100dB SNR 2.5MS/s output data rate ΔΣ ADC. In: IEEE International Solid-State Circuits Conference (ISSCC) Digest Technical Papers, vol 1, pp 172–591

M. Figueiredo et al., *Reference-Free CMOS Pipeline Analog-to-Digital Converters*, Analog Circuits and Signal Processing, DOI: 10.1007/978-1-4614-3467-2,

15. Brooks L, Lee HS (2007) A zero-crossing-based 8-bit 200 MS/s pipelined ADC. IEEE J Solid-State Circuits 42(12):2677–2687
16. Brooks T, Robertson D, Kelly D, Del Muro A, Harston S (1997) A cascaded sigma-delta pipeline A/D converter with 1.25 MHz signal bandwidth and 89 dB SNR. IEEE J Solid-State Circuits 32(12):1896–1906
17. Bult K (2009) Embedded analog-to-digital converters. In: Proceedings of European Solid-State Circuits Conference (ESSCIRC), pp 52–64
18. Callewaert L, Sansen W (1990) Class AB CMOS amplifiers with high efficiency. IEEE J Solid-State Circuits 25(3):684–691
19. Cao Z, Song T, Yan S (2007) A 14 mW 2.5 MS/s 14 bit ΣΔ modulator using split-path pseudo-differential amplifiers. IEEE J Solid-State Circuits 42(10):2169–2179
20. Cao Z, Yan S, Li Y (2009) A 32 mW 1.25 GS/s 6b 2b/Step SAR ADC in 0.13μm CMOS. IEEE J Solid-State Circuits 44(3):862–873
21. Chandrashekar K, Bakkaloglu B (2011) A 10 b 50 MS/s opamp-sharing pipeline A/D with current-reuse OTAs. IEEE Trans VLSI Syst 19(9):1610–1616
22. Chang TH, Dung LR (2008) Fourth-Order cascaded ΣΔ modulator using tri-level quantization and bandpass noise shaping for broadband telecommunication applications. IEEE Trans Circuits Syst I, Reg Papers 55(6):1722–1732
23. Chang TH, Dung LR (2008) Fourth-order cascaded ΣΔ modulator using tri-level quantization and bandpass noise shaping for broadband telecommunication applications. IEEE Trans Circuits Syst I, Reg Papers 55(6):1722–1732
24. Chang TH, Dung LR, Guo JY, Yang KJ (2007) A 2.5-V 14-bit, 180-mW cascaded ΣΔ ADC for ADSL2+ application. IEEE J Solid-State Circuits 42(11):2357–2368
25. Chen HS, Song BS, Bacrania K (2001) A 14-b 20-Msamples/s CMOS pipelined ADC. IEEE J Solid-State Circuits 36(6):997–1001
26. Chen HW, Chen IC, Tseng HC, Chen HS (2009) A 1-GS/s 6-bit two-channel two-step ADC in 0.13-μm CMOS. IEEE J Solid-State Circuits 44(11):3051–3059
27. Chew K, Yeo K, Chu SF (2004) Effect of technology scaling on the 1/f noise of deep submicron PMOS transistors. Elsevier Solid-State Electronics 48(7):1101–1109
28. Chilakapati U, Fiez T (1998) Settling time design considerations for SC integrators. In: Proceedings of IEEE International Symposium Circuits Systems (ISCAS), vol 1, pp 492–495
29. Chiu Y (2000) Inherently linear capacitor error-averaging techniques for pipelined A/D conversion. IEEE Trans Circuits Syst II 47(3):229–232
30. Chiu Y, Gray PR, Nikolic B (2004) A 14-b 12-MS/s CMOS pipeline ADC with over 100-dB sfdr. IEEE J Solid-State Circuits 39(12):2139–2151
31. Cho YJ, Lee SH (2005) An 11b 70-MHz 1.2-mm^2 49-mW 0.18-μm CMOS ADC with on-chip current/voltage references. IEEE Trans Circuits Syst I, Reg Papers 52(10):1989–1995
32. Cho YJ, Bae HH, Lee MJ, Kim MJ, Lee SH, Kim YL (2004) An 8b 220 MS/s 0.25 μm CMOS pipeline ADC with on-chip RC-filter based voltage references. In: Proceedings of IEEE Asia-Pacific Conference Advanced Systems Integrated Circuits, pp 90–93
33. Cho YK, Jeon YD, Nam JW, Kwon JK (2010) A 9-bit 80 MS/s successive approximation register analog-to-digital converter with a capacitor reduction technique. IEEE Trans Circuits Syst II, Exp Briefs 57(7):502–506
34. Choi HC, Kim YJ, Yoo SW, Hwang SY, Lee SH (2008) A programmable 0.8-V 10-bit 60-MS/s 192-mW 013-μm CMOS ADC operating down to 0 5 V. IEEE Trans Circuits Syst II, Exp Briefs 55(4):319–323
35. Choksi O, Carley L (2003) Analysis of switched-capacitor common-mode feedback circuit. IEEE Trans Circuits Syst II 50(12):906–917
36. Christen T, Huang Q (2010) A 0.13 μm CMOS 0.1-20MHz bandwidth 86-70dB DR multi-mode DT ΔΣ ADC for imt-advanced. In: Proceedings of European Solid-State Circuits Conference (ESSCIRC), pp 414–417
37. Conroy CSG, Cline DW, Gray PR (1993) An 8-b 85-MS/s parallel pipeline A/D converter in 1-μm CMOS. IEEE J Solid-State Circuits 28(4):447–454

38. Custódio J, Oliveira L, Goes J, Oliveira J, Bruun E, Andreani P (2010) A small-area self-biased wideband CMOS balun LNA with noise cancelling and gain enhancement. In: NORCHIP, pp 1–4
39. Daihong F, Dyer KC, Lewis SH, Hurst PJ (1998) A digital background calibration technique for time-interleaved analog-to-digital converters. IEEE J Solid-State Circuits 33(12):1904–1911
40. del Rio R, de la Rosa J, Perez-Verdu B, Delgado-Restituto M, Dominguez-Castro R, Medeiro F, Rodriguez-Vazquez A (2004) Highly linear 2.5-V CMOS ΣΔ modulator for ADSL+. IEEE Trans Circuits Syst I, Reg Papers 51(1):47–62
41. Dessouky M, Kaiser A (1999) Input switch configuration suitable for rail-to-rail operation of switched op amp circuits. IET Electronics Letters 35(1):8–10
42. Doernberg J, Lee HS, Hodges DA (1984) Full-speed testing of A/D converters. IEEE J Solid-State Circuits 19(6):820–827
43. Eschauzier R, Kerklaan L, Huijsing J (1992) A 100-MHz 100-dB operational amplifier with multipath nested Miller compensation structure. IEEE J Solid-State Circuits 27(12):1709–1717
44. Ferri G, Sansen W (1997) A rail-to-rail constant-gm low-voltage CMOS operational transconductance amplifier. IEEE J Solid-State Circuits 32(10):1563–1567
45. Fiez T, Yang H, Yang J, Yu C, Allstot D (1989) A family of high-swing CMOS operational amplifiers. IEEE J Solid-State Circuits 24(6):1683–1687
46. Figueiredo M, Michalak T, Goes J, Gomes L, Sniatala P (2009) Improved clock-phase generator based on self-biased CMOS logic for time-interleaved SC circuits. In: Proceedings of IEEE International Conference on Electronics, Circuits, and Systems (ICECS), pp 763–766
47. Figueiredo M, Santin E, Goes J, Santos-Tavares R, Evans G (2010) Two-stage fully-differential inverter-based self-biased CMOS amplifier with high efficiency. In: Proceedings of IEEE International Symposium on Circuits and Systems (ISCAS), pp 2828–2831
48. Figueiredo M, Goes J, Oliveira LB, Steiger-Garção A (2011a) Low voltage low power fully differential self-biased 1.5-bit quantizer with built-in thresholds. Int J of Circuit Theory and Applications Early View
49. Figueiredo M, Santin E, Goes J, Paulino N, Barúqui F, Petraglia A (2011) Flipped-around multiply-by-two amplifier with unity feedback factor. Analog Integr Circuits Process 68(1):133–138
50. Figueiredo M, Santos-Tavares R, Santin E, Ferreira J, Evans G, Goes J (2011) A two-stage fully differential inverter-based self-biased CMOS amplifier with high efficiency. IEEE Trans Circuits Syst I, Reg Papers 58(7):1591–1603
51. Figueiredo M, Santin E, Goes J, Evans G, Paulino N (2012) A reference-free 7-bit 500-MS/s pipeline ADC using current-mode reference shifting and built-in threshold quantizers. submitted for publication in Analog Integr Circuits Process
52. Figueiredo PM, Vital JC (2009) Offset reduction techniques in high-speed analog-to-digital converters: analysis, design and tradeoffs. Springer, Dordrecht ; London
53. Fujimori I, Longo L, Hairapetian A, Seiyama K, Kosic S, Jun C, Shu-Lap C (2000) A 90-dB SNR 2.5-MHz output-rate ADC using cascaded multibit delta-sigma modulation at 8× oversampling ratio. IEEE J Solid-State Circuits 35(12):1820–1828
54. Gaggl R, Inversi M, Wiesbauer A (2004) A power optimized 14-bit SC ΔΣ modulator for ADSL CO applications. In: IEEE International Solid-State Circuits Conference (ISSCC) Digest Technical Papers, vol 1, pp 82–514
55. Giustolisi G, Grasso AD, Pennisi S (2007) High-drive and linear CMOS class-AB pseudo-differential amplifier. IEEE Trans Circuits Syst II, Exp Briefs 54(2):112–116
56. Goes J, Pereira J, Paulino N, Silva MM (2007) Switched-capacitor multiply-by-two amplifier insensitive to component mismatches. IEEE Trans Circuits Syst II, Exp Briefs 54(1):29–33

57. Goll B, Zimmermann H (2005) A low-power 2-GSample/s comparator in 120 nm CMOS technology. In: Proceedings of European Solid-State Circuits Conference (ESSCIRC), pp 507–510
58. Goll B., Zimmermann H. (2006) A low-power 4GHz comparator in 120nm CMOS technology with a technique to tune resolution. In: Proceedings of European Solid-State Circuits Conference (ESSCIRC), pp 320–323
59. Goll B, Zimmermann H (2007a) A 0.12 μm CMOS comparator requiring 0.5V at 600MHz and 1.5V at 6GHz. In: IEEE International Solid-State Circuits Conference (ISSCC) Digest Technical Papers, pp 316–605
60. Goll B, Zimmermann H (2007b) A clocked, regenerative comparator in 0.12 μm CMOS with tunable sensitivity. In: Proceedings of European Solid-State Circuits Conference (ESSCIRC), pp 408–411
61. Goll B, Zimmermann H (2009a) A 65nm CMOS comparator with modified latch to achieve 7GHz/1.3mW at 1.2V and 700MHz/47 μw at 0.6V. In: IEEE International Solid-State Circuits Conference (ISSCC) Digest Technical Papers, pp 328–329,329a
62. Goll B, Zimmermann H (2009) A comparator with reduced delay time in 65-nm CMOS for supply voltages down to 0.65 V. IEEE Trans Circuits Syst II, Exp Briefs 56(11):810–814
63. Grasso A, Palumbo G, Pennisi S (2007) Advances in reversed nested Miller compensation. IEEE Trans Circuits Syst I, Reg Papers 54(7):1459–1470
64. Grasso A, Palumbo G, Pennisi S (2008) Comparison of the frequency compensation techniques for CMOS two-stage Miller otas. IEEE Trans Circuits Syst II, Exp Briefs 55(11):1099–1103
65. Gray PR, Hurst PJ, Lewis SH, Meyer RG (2009) Analysis and design of analog integrated circuits, 5th edn. Wiley, New York
66. Gregorian R, Temes GC (1986) Analog MOS integrated circuits for signal processing. Wiley, New York ; Chichester
67. Guilherme J.M.C. (2003) Architectures for high dynamic range CMOS pipeline analogue-to-digital convertion. PhD thesis, Universidade Técnica de Lisboa
68. Gulati K, Munoz C, Seonghwan C, Manganaro G, Lugin M, Peng M, Pulincherry A, Jipeng L, Bugeja A, Chandrakasan A, Shoemaker D (2004) A highly integrated analog baseband transceiver featuring a 12-bit 180MSPS pipelined A/D converter for multi-channel wireless LAN. In: Symposium on VLSI Circuits Digest of Technical Papers, pp 428–431
69. Gulati K, Peng MS, Pulincherry A, Munoz CE, Lugin M, Bugeja AR, Jipeng L, Chandrakasan AP (2006) A highly integrated CMOS analog baseband transceiver with 180 MSPS 13-bit pipelined CMOS ADC and dual 12-bit DACs. IEEE J Solid-State Circuits 41(8):1856–1866
70. Gupta S, Fong V (2002) A 64-MHz clock-rate $\Sigma\Delta$ ADC with 88-dB sndr and-105-dB im3 distortion at a 1.5-MHz signal frequency. IEEE J Solid-State Circuits 37(12):1653–1661
71. Hajimiri A (2010) Generalized time- and transfer-constant circuit analysis. IEEE Trans Circuits Syst I, Reg Papers 57(6):1105–1121
72. He C, Chen D, Geiger R (2003) A low-voltage compatible two-stage amplifier with $\geq$120 dB gain in standard digital CMOS. In: Proceedings of IEEE International Symposium on Circuits and Systems (ISCAS), vol 1, pp 353–356
73. He XY, Pun KP, Kinget P (2009) A 0.5-V wideband amplifier for a 1-MHz CT complex delta-sigma modulator. IEEE Trans Circuits Syst II, Exp Briefs 56(11):805–809
74. Hogervorst R, Tero J, Eschauzier R, Huijsing J (1994) A compact power-efficient 3 V CMOS rail-to-rail input/output operational amplifier for vlsi cell libraries. IEEE J Solid-State Circuits 29(12):1505–1513
75. Hsu CC, Huang CC, Lin YH, Lee CC, Soe Z, Aytur T, Yan RH (2007) A 7b 1.1GS/s reconfigurable time-interleaved ADC in 90nm CMOS. In: Symposium on VLSI Circuits Digest of Technical Papers pp 66–67

76. Hu J, Dolev N, Murmann B (2009) A 9.4-bit, 50-MS/s, 1.44-mW pipelined ADC using dynamic source follower residue amplification. IEEE J Solid-State Circuits 44(4):1057–1066
77. Inc AD (2004) Analog-Digital Conversion. Analog Devices Inc., USA
78. Ishii H, Tanabe K, Iida T (2005) A 1.0 V 40mW 10b 100MS/s pipeline ADC in 90nm CMOS. In: Proceedings of IEEE Custom Integrated Circuits Conference (CICC), pp 395–398
79. Limotyrakis J.K., Yang S. CKK (2011) Multilevel power optimization of pipelined A/D converters. IEEE Trans VLSI Syst 19(5):832–845
80. Jenq YC (1988) Digital spectra of nonuniformly sampled signals: fundamentals and high-speed waveform digitizers. IEEE Trans Instrum Meas 37(2):245–251
81. Jeon YD, Cho YK, Nam JW, Kim KD, Lee WY, Hong KT, Kwon JK (2010) A 9.15mW 0.22mm^2 10b 204MS/s pipelined SAR ADC in 65nm CMOS. In: Proceedings of IEEE Custom Integrated Circuits Conference (CICC), pp 1–4
82. Jeong DK, Borriello G, Hodges DA, Katz RH (1987) Design of PLL-based clock generation circuits. IEEE J Solid-State Circuits 22(2):255–261
83. Jiang S, Do MA, Yeo KS, Lim WM (2008) An 8-bit 200-MSample/s pipelined ADC with mixed-mode front-end S/H circuit. IEEE Trans Circuits Syst I, Reg Papers 55(6):1430–1440
84. Johns D, Martin KW (1997) Analog integrated circuit design. Wiley, New York
85. Jussila J, Ryynanen J, Kivekas K, Sumanen L, Parssinen A, Halonen K (2001) A 22-mA 3.0-dB NF direct conversion receiver for 3G WCDMA. IEEE J Solid-State Circuits 36(12):2025–2029
86. Karanicolas AN, Lee HS, Barcrania KL (1993) A 15-b 1-Msample/s digitally self-calibrated pipeline ADC. IEEE J Solid-State Circuits 28(12):1207–1215
87. Keramat A., Tao Z. (2000) A capacitor mismatch and gain insensitive 1.5-bit/stage pipelined A/D converter. In: Proceedings of IEEE Midwest Symposium Circuits Systems, vol 1, pp 48–51
88. Khanoyan K, Behbahani F, Abidi AA (1999) A 10 b, 400 MS/s glitch-free CMOS D/A converter. In: Symposium on VLSI Circuits Digest of Technical Papers, pp 73–76
89. Khorram S, Darabi H, Zhou Z, Li Q, Marholev B, Chiu J, Castaneda J, Chien HM, Anand S, Wu S, Pan MA, Roofougaran R, Kim HJ, Lettieri P, Ibrahim B, Rael J, Tran L, Geronaga E, Yeh H, Frost T, Trachewsky J, Rofougaran A (2005) A fully integrated SOC for 802.11b in 0.18-μm CMOS. IEEE J Solid-State Circuits 40(12):2492–2501
90. Kim HC, Jeong DK, Kim W (2005) A 30mW 8b 200MS/s pipelined CMOS ADC using a switched-opamp technique. In: IEEE International Solid-State Circuits Conference (ISSCC) Digest Technical Papers, vol 1, pp 284–598
91. Kurosawa N, Kobayashi H, Maruyama K, Sugawara H, Kobayashi K (2001) Explicit analysis of channel mismatch effects in time-interleaved ADC systems. IEEE Trans Circuits Syst I 48(3):261–271
92. Laker KR, Sansen WMC (1994) Design of analog integrated circuits and systems. McGraw-Hill, New York ; London
93. Lakshmikumar K, Hadaway R, Copeland M (1986) Characterisation and modeling of mismatch in MOS transistors for precision analog design. IEEE J Solid-State Circuits 21(6):1057–1066
94. Lee D, Yoo J, Choi K (2002) Design method and automation of comparator generation for flash A/D converter. In: Proceedings of International Symposium Quality Electronic Design, pp 138–142
95. Lee KS, Choi Y, Maloberti F (2005) SC amplifier and SC integrator with an accurate gain of 2. IEEE Trans Circuits Syst II, Exp Briefs 52(4):194–198
96. Leung KN, Mok P (2001) Analysis of multistage amplifier-frequency compensation. IEEE Trans Circuits Syst I 48(9):1041–1056
97. Lewis SH, Gray PR (1987) A pipelined 5-Msample/s 9-bit analog-to-digital converter. IEEE J Solid-State Circuits 22(6):954–961

98. Lewis SH, Fetterman HS, Gross GF Jr, Ramachandran R, Viswanathan TR (1992) A 10-b 20-Msample/s analog-to-digital converter. IEEE J Solid-State Circuits 27(3):351–358
99. Leyn F, Daems W, Gielen G, Sansen W (1997) A behavioral signal path modeling methodology for qualitative insight in and efficient sizing of CMOS opamps. In: IEEE/ACM International Conference on Computer-Aided Design Digested Technical Papers, pp 374–381
100. Li Y, Sanchez-Sinencio E (2003) A wide input bandwidth 7-bit 300-MSample/s folding and current-mode interpolating ADC. IEEE J Solid-State Circuits 38(8):1405–1410
101. Lin D, Li L, Farahani S, Flynn M (2010) A flexible 500MHz to 3.6GHz wireless receiver with configurable DT FIR and IIR filter embedded in a 7b 21MS/s SAR ADC. In: Proceedings of IEEE Custom Integrated Circuits Conference (CICC), pp 1–4
102. Lin YM, Kim B, Gray PR (1991) A 13-b 2.5-MHz self-calibrated pipelined A/D converter in 3-μm CMOS. IEEE J Solid-State Circuits 26(4):628–636
103. Macq D, Jespers PGA (1994) A 10-bit pipelined switched-current A/D converter. IEEE J Solid-State Circuits 29(8):967–971
104. Maloberti F (2001) Analog design for CMOS VLSI systems. Kluwer Academic, Boston
105. Maloberti F (2007) Data converters. Springer, Dordrecht
106. Mandal P, Visvanathan V (1997) A self-biased high performance folded cascode CMOS op-amp. In: Proceedings of Tenth International Conference VLSI Design, pp 429–434
107. Manghisoni M, Ratti L, Re V, Speziali V, Traversi G (2006) Noise characterization of 130 nm and 90 nm CMOS technologies for analog front-end electronics. In: Conference Record IEEE Nuclear Science Symposium, pp 214–218
108. ATLAB (2011) The MATLAB website. http://www.mathworks.com/
109. Maulik PC, Chadha MS, Lee WL, Crawley PJ (2000) A 16-bit 250-kHz delta-sigma modulator and decimation filter. IEEE J Solid-State Circuits 35(4):458–467
110. MAX (2011) Maxim MAX1121. http://www.maxim-ic.com/datasheet/index.mvp/id/4156
111. Miyahara M, Matsuzawa A (2009) A low-offset latched comparator using zero-static power dynamic offset cancellation technique. In: IEEE Asian Solid-State Circuits Conf. (ASSCC), pp 233–236
112. Miyahara M, Asada Y, Paik D, Matsuzawa A (2008) A low-noise self-calibrating dynamic comparator for high-speed ADCs. In: IEEE Asian Solid-State Circuits Conference (ASSCC), pp 269–272
113. Mok WI, Mak PI, U SP, Martins RP (2004) Modeling of noise sources in reference voltage generator for very-high-speed pipelined ADC. In: Proceedings of IEEE Midwest Symposium Circuits Systems, vol 1, pp I–5–8
114. Murmann B, Boser BE (2003) A 12-bit 75-MS/s pipelined ADC using open-loop residue amplification. IEEE J Solid-State Circuits 38(12):2040–2050
115. Nazemi A, Grace C, Lewyn L, Kobeissy B, Agazzi O, Voois P, Abidin C, Eaton G, Kargar M, Marquez C, Ramprasad S, Bollo F, Posse V, Wang S, Asmanis G (2008) A 10.3GS/s 6bit (5.1 ENOB at nyquist) time-interleaved/pipelined ADC using open-loop amplifiers and digital calibration in 90nm CMOS. In: Symposium VLSI Circuits Digest Technical Papers, pp 18–19
116. Nedovic N, Walker WW, Oklobdzija VG, Aleksic M (2002) A low power symmetrically pulsed dual edge-triggered flip-flop. In: Proceedings of European Solid State Circuits Conference (ESSCIRC), pp 399–402
117. Nezuka T, Misawa K, Azami J, Majima Y, Okamura JI (2006) A 10-bit 200MS/s pipeline A/D Converter for high-speed video signal digitizer. In: IEEE Asian Solid-State Circuits Conference (ASSCC), pp 31–34
118. Ng HT, Ziazadeh R, Allstot D (1999) A multistage amplifier technique with embedded frequency compensation. IEEE J Solid-State Circuits 34(3):339–347
119. Nguyen R, Murmann B (2010) The design of fast-settling three-stage amplifiers using the open-loop damping factor as a design parameter. IEEE Trans Circuits Syst I, Reg Papers 57(6):1244–1254

120. Okamoto K, Morie T, Yamamoto A, Nagano K, Sushihara K, Nakahira H, Horibe R, Aida K, Takahashi T, Ochiai M, Soneda A, Kakiage T, Iwasaki T, Taniuchi H, Shibata T, Ochi T, Takiguchi M, Yamamoto T, Seike T, Matsuzawa A (2003) A fully integrated 0.13-μm CMOS mixed-signal SoC for DVD player applications. IEEE J Solid-State Circuits 38(11):1981–1991
121. Okaniwa Y, Tamura H, Kibune M, Yamazaki D, Cheung TS, Ogawa J, Tzartzanis N, Walker W, Kuroda T (2005) A 40-Gb/s CMOS clocked comparator with bandwidth modulation technique. IEEE J Solid-State Circuits 40(8):1680–1687
122. Oliveira J, Goes J, Esperança B, Paulino N, Fernandes J (2007) Low-power CMOS comparator with embedded amplification for ultra-high-speed ADCs. In: Proceedings of IEEE International Symposium on Circuits and Systems (ISCAS), pp 3602–3605
123. Oliveira J, Goes J, Figueiredo M, Santin E, Fernandes J, Ferreira J (2010) An 8-bit 120-MS/s interleaved CMOS pipeline ADC based on MOS parametric amplification. IEEE Trans Circuits Syst II, Exp Briefs 57(2):105–109
124. Op't Eynde F, Ampe P, Verdeyen L, Sansen W (1990) A CMOS large-swing low-distortion three-stage class AB power amplifier. IEEE J Solid-State Circuits 25(1):265–273
125. Paavola M, Kamarainen M, Jarvinen J, Saukoski M, Laiho M, Halonen K (2007) A micropower interface ASIC for a capacitive 3-axis micro-accelerometer. IEEE J Solid-State Circuits 42(12):2651–2665
126. Paavola M, Kamarainen M, Laulainen E, Saukoski M, Koskinen L, Kosunen M, Halonen K (2009) A micropower-based interface ASIC for a capacitive 3-axis micro-accelerometer. IEEE J Solid-State Circuits 44(11):3193–3210
127. Pacheco J, Figueiredo M, Paulino N, Goes J (2012) Current-mode reference shifting solution for MDAC-based analog-to-digital converters. In: accepted for publication in Proceedings of IEEE International Symposium on Circuits and Systems (ISCAS)
128. Park S, Flynn M (2006) A regenerative comparator structure with integrated inductors. IEEE Trans Circuits Syst I, Reg Papers 53(8):1704–1711
129. Park S.M., J-Band Yoo, Kim S.W., Cho Y.J., Lee S.H. (2004) A 10-b 150-MSample/s 1.8-V 123-mW CMOS A/D converter with 400-MHz input bandwidth. IEEE J Solid-State Circuits 39(8):1335–1337
130. Pei S, Chan SP (1990) DSP application in interal-non-linearity testing of A/D converters. In: Conference Record Asilomar Conference on Signals and Computers, vol 1 p 516
131. Pelgrom M, Duinmaijer A, Welbers A (1989) Matching properties of MOS transistors. IEEE J Solid-State Circuits 24(5):1433–1439
132. Perez A, Nithin K, Bonizzoni E, Maloberti F (2009) Slew-rate and gain enhancement in two stage operational amplifiers. In: Proceedings of IEEE International Symposium on Circuits and Systems (ISCAS), pp 2485–2488
133. Petraglia A, Mitra SK (1991) Analysis of mismatch effects among A/D converters in a time-interleaved waveform digitizer. IEEE Trans Instrum Meas 40(5):831–835
134. Picolli L, Malcovati P, Crespi L, Chaahoub F, Baschirotto A (2008) A 90nm 8b 120MS/s-250MS/s pipeline ADC. In: Proceedings of European Solid State Circuits Conference (ESSCIRC), pp 266–269
135. Plassche RJ vd (1994) Integrated analog-to-digital and digital-to-analog converters. Kluwer, Boston ; London
136. Poulton K, Neff R, Muto A, Liu W, Burstein A, Heshami M (2002) A 4GSample/s 8b ADC in 0.35 μm CMOS. In: IEEE International Solid-State Circuits Conference (ISSCC) Digest Technical Papers, vol 2, pp 126–434
137. Purcell J, Abdel-Aty-Zohdy H (1997) Compact high gain CMOS op amp design using comparators. In: Proceedings of IEEE Midwest Symposim Circuits Systems, vol 2, pp 1050–1052
138. Quinn P, Pribytko M (2003) Capacitor matching insensitive 12-bit 3.3 MS/s algorithmic ADC in 0.25 μm CMOS. In: Proceedings of IEEE Custom Integrated Circuits Conference (CICC), pp 425–428

139. Ramesh J, Gunavathi K (2007) A 8-bit TIQ based 780 MSPS CMOS flash A/D converter. In: International Conference Computational Intelligence and Multimedia Applications, vol 2, pp 201–208
140. Razavi B (1995) Principles of data conversion system design. IEEE Press, New York
141. Razavi B (2001) Design of analog CMOS integrated circuits. McGraw-Hill, Boston
142. Real P, Robertson DH, Mangelsdorf CW, Tewksbury TL (1991) A wide-band 10-b 20 MS/s pipelined ADC using current-mode signals. IEEE J Solid-State Circuits 26(8):1103–1109
143. Ruiz-Amaya J, Delgado-Restituto M, Rodriguez-Vazquez A (2009) Accurate settling-time modeling and design procedures for two-stage Miller-compensated amplifiers for switched-capacitor circuits. IEEE Trans Circuits Syst I, Reg Papers 56(6):1077–1087
144. Saberi M, Lotfi R (2007) A capacitor mismatch-and nonlinearity-insensitive 1.5-bit residue stage for pipelined ADCs. In: Proceedings of IEEE International Conference on Electronics, Circuits and Systems (ICECS), pp 677–680
145. Sansen W, Chang Z (1990) Feedforward compensation techniques for high-frequency CMOS amplifiers. IEEE J Solid-State Circuits 25(6):1590–1595
146. Sansen WMC (2006) Analog design essentials. Springer-Verlag New York, Inc.
147. Santin E, Figueiredo M, Tavares R, Goes J, Oliveira L (2010) Fast-settling low-power two-stage self-biased CMOS amplifier using feedforward-regulated cascode devices. In: Proceedings of IEEE International Conference on Electronics, Circuits and Systems (ICECS), pp 25–28
148. Schinkel D, Mensink E, Klumperink E, van Tuijl E, Nauta B (2007) A double-tail latch-type voltage sense amplifier with 18ps setup+hold time. In: IEEE International Solid-State Circuits Conference (ISSCC) Digest Technical Papers, pp 314–605
149. Schlogl F, Zimmermann H (2004) 1.5 GHz opamp in 120nm digital CMOS. In: Proceedings of European Solid-State Circuits Conference (ESSCIRC), pp 239–242
150. Schreier R, Silva J, Steensgaard J, Temes GC (2005) Design-oriented estimation of thermal noise in switched-capacitor circuits. IEEE Trans Circuits Syst I, Reg Papers 52(11):2358–2368
151. Sepke T, Holloway P, Sodini C, Lee HS (2009) Noise analysis for comparator-based circuits. IEEE Trans Circuits Syst I, Reg Papers 56(3):541–553
152. Shankar A, Silva-Martinez J, Sanchez-Sinencio E (2001) A low voltage operational transconductance amplifier using common mode feedforward for high frequency switched capacitor circuits. In: Proceedings IEEE International Symposium Circuits Systems (ISCAS), pp 643–646
153. Sheikhaei S, Mirabbasi S, Ivanov A (2005) A 0.35 μm CMOS comparator circuit for high-speed ADC applications. In: Proceedings IEEE Innternational Symposium Circuits Systems (ISCAS), vol 6, pp 6134–6137
154. Silva A, Cavalheiro D, Abdollahvand S, Oliveira L, Figueiredo M, Goes J (2011) A self-biased ring oscillator with quadrature outputs operating at 600 MHz in a 130 nm CMOS technology. In: Proceedings International Conference Mixed Design of Integrated Circuits Systems (MIXDES), pp 221–224
155. Singer L, Ho S, Timko M, Kelly D (2000) A 12 b 65 MSample/s CMOS ADC with 82 dB SFDR at 120 MHz. In: IEEE International Solid-State Circuits Conference (ISSCC) Digest of Technical Papers, pp 38–39
156. Song BS, Tompsett MF, Lakshmikumar KR (1988) A 12-bit 1-Msample/s capacitor error-averaging pipelined A/D converter. IEEE J Solid-State Circuits 23(6):1324–1333
157. Song WC, Choi HW, Kwak SU, Song BS (1995) A 10-b 20-Msample/s low-power CMOS ADC. IEEE J Solid-State Circuits 30(5):514–521
158. Sumanen L, Waltari M, Halonen KAI (2001) A 10-bit 200-MS/s CMOS parallel pipeline A/D converter. IEEE J Solid-State Circuits 36(7):1048–1055
159. Taherzadeh-Sani M, Hamoui A (2011) A 1-V process-insensitive current-scalable two-stage opamp with enhanced dc gain and settling behavior in 65-nm digital CMOS. IEEE J Solid-State Circuits 46(3):660–668

160. Taherzadeh-Sani M, Lofti R, Shoaei O (2003) A novel frequency compensation technique for two-stage CMOS operational amplifiers. In: Proceedings IEEEInternational Conference on Electronics, Circuits and Systems (ICECS), pp 256–259
161. Tavares RMLS (2010) Time-domain optimization of amplifiers based on distributed genetic algorithms. PhD thesis, Universidade Nova de Lisboa
162. Thandri B, Silva-Martinez J (2003) A robust feedforward compensation scheme for multistage operational transconductance amplifiers with no Miller capacitors. IEEE J Solid-State Circuits 38(2):237–243
163. Thoutam S, Ramirez-Angulo J, Lopez-Martin A, Carvajal R (2004) Power efficient fully differential low-voltage two stage class AB/AB op-amp architectures. In: Proceedings IEEE International Symposium on Circuits and Systems (ISCAS), vol 1, pp 733–736
164. TI (2011) Texas Instruments THS4521D. http://focus.ti.com/docs/prod/folders/print/ths4521.html
165. Torelli G, DeLa Plaza A (1998) Tracking switched-capacitor CMOS current reference. IEE Proceedings Circuits, Devices Systems 145(1):44–47
166. Trojer M, Cleris M, Gaier U, Hebein T, Pridnig P, Kuttin B, Tschuden B, Krassnitzer C, Kuttin C, Pribyl W (2008) A 1.2V 56mW 10 bit 165MS/s pipeline-ADC for HD-video applications. In: Proceedings of European Solid-State Circuits Confernce (ESSCIRC), pp 270–273
167. Tsividis Y (1999) Operation and modeling of the MOS transistor. Oxford University Press, New York
168. Tu WH, Kang TH (2008) A 1.2V 30mW 8b 800MS/s time-interleaved ADC in 65nm CMOS. In: Symposium VLSI Circuits Digest Technical Papers, pp 72–73
169. Varzaghani A, Yang CK (2006) A 600-MS/s 5-bit pipeline A/D converter using digital reference calibration. IEEE J Solid-State Circuits 41(2):310–319
170. Vaz BMNJ (2005) Conversores analógico-digital de elevada velocidade e tensão de alimentação reduzida. PhD thesis, Universidade Nova de Lisboa
171. Verma A, Razavi B (2009) A 10-Bit 500-MS/s 55-mW CMOS ADC. IEEE J Solid-State Circuits 44(11):3039–3050
172. Vogel C (2005) The impact of combined channel mismatch effects in time-interleaved ADCs. IEEE Trans Instrum Meas 54(1):415–427
173. Walden R (1994) Analog-to-digital converter technology comparison. In: 16th Annual GaAs IC Symposium Technical Digest, pp 217–219
174. Wang S, Niknejad A, Brodersen R (2006) Design of a sub-mW 960-MHz UWB CMOS LNA. IEEE J Solid-State Circuits 41(11):2449–2456
175. Wong KL, Yang CK (2004) Offset compensation in comparators with minimum input-referred supply noise. IEEE J Solid-State Circuits 39(5):837–840
176. Wong YL, Cohen M, Abshire P (2008) A 1.2-GHz comparator with adaptable offset in 0.35-μm CMOS. IEEE Trans Circuits Syst I, Reg Papers 55(9):2584–2594
177. Wu JT, Wooley B (1988) A 100-MHz pipelined CMOS comparator. IEEE J Solid-State Circuits 23(6):1379–1385
178. Yamamoto T, Gotoh SI, Takahashi T, Irie K, Ohshima K, Mimura N, Aida K, Maeda T, Sushihara K, Okamoto Y, Tai Y, Usui M, Nakajima T, Ochi T, Komichi K, Matsuzawa A (2001) A mixed-signal 0.18-μm CMOS SoC for DVD systems with 432-MSample/s PRML read channel and 16-Mb embedded DRAM. IEEE J Solid-State Circuits 36(11):1785–1794
179. Yan J, Geiger R (2002) A high gain CMOS operational amplifier with negative conductance gain enhancement. In: Proceedings of IEEE Custom Integrated Circuits Conference (CICC), pp 337–340
180. Yang CY, Liu SI (2001) A one-wire approach for skew-compensating clock distribution based on bidirectional techniques. IEEE J Solid-State Circuits 36(2):266–272
181. Yang W, Kelly D, Mehr L, Sayuk M, Singer L (2001) A 3-V 340-mW 14-b 75-Msample/s CMOS ADC with 85-dB SFDR at nyquist input. IEEE J Solid-State Circuits 36(12):1931–1936

182. Yao L, Steyaert M, Sansen W (2002) Fast-settling CMOS two-stage operational transconductance amplifiers and their systematic design. In: Proceedings of IEEE International Symposium Circuits Systems (ISCAS), vol 2, pp 839–842
183. Yavari M, Zare-Hoseini H, Farazian M, Shoaei O (2003) A new compensation technique for two-stage CMOS operational transconductance amplifiers. In: Proceedings of IEEE International Conference Electronics, Circuits, Systems (ICECS), vol 2, pp 539–542
184. Yavari M, Shoaei O, Svelto F (2005) Hybrid cascode compensation for two-stage CMOS operational amplifiers. In: Proceedings of IEEE International Symposium Circuits Systems (ISCAS), vol 2, pp 1565–1568
185. Yoo J, Lee D, Choi K, Tangel A (2001) Future-ready ultrafast 8bit CMOS ADC for system-on-chip applications. In: Proceedings of Annual IEEE International ASIC/SOC Conference, pp 455–459
186. Yoo J, Choi K, Ghaznavi J (2003) Quantum voltage comparator for 0.07 μm CMOS flash A/D converters. In: Proceedings IEEE Computer Society Annual Symposium VLSI, pp 280–281
187. Yu H, Chin S, Wong B (2008) A 12b 50MSPS 34mW pipelined ADC. In: Proceedings IEEE Custom Integrated Circuits Conference (CICC), pp 297–300
188. Zare-Hoseini H, Shoaei O, Kale I (2005) Multiply-by-two gain stage with reduced mismatch sensitivity. IET Electronics Letters 41(6):289–290
189. Zare-Hoseini H, Shoaei O, Kale I (2006) A new structure for capacitor-mismatch-insensitive multiply-by-two amplification. In: Proceedings of IEEE nternational Symposium Circuits Systems (ISCAS), pp 4879–4882
190. Zhian Tabasy E, Kamarei M, Ashtiani S (2009) 1.5-bit mismatch-insensitive MDAC with reduced input capacitive loading. IET Electronics Letters 45(23):1157–1158
191. Zojer B, Petschacher R, Luschnig W (1985) A 6-bit/200-MHz full nyquist A/D converter. IEEE J Solid-State Circuits 20(3):780–786

Index

M. Figueiredo et al., *Reference-Free CMOS Pipeline Analog-to-Digital Converters*,
Analog Circuits and Signal Processing, DOI: 10.1007/978-1-4614-3467-2,